THE MATHEMATICAL AND PHYSICAL MODELING OF PRIMARY METALS PROCESSING OPERATIONS

THE MATHEMATICAL AND PHYSICAL MODELING OF PRIMARY METALS PROCESSING OPERATIONS

JULIAN SZEKELY

Massachusetts Institute of Technology
Cambridge, Massachusetts

JAMES W. EVANS

University of California, Berkeley
Berkeley, California

J. K. BRIMACOMBE

University of British Columbia
Vancouver, British Columbia

A Wiley-Interscience Publication

JOHN WILEY & SONS

New York / Chichester / Brisbane / Toronto / Singapore

"To the Memory of
Professor T. B. King"

Library of Congress Cataloging-in-Publication Data:

Szekely, Julian, 1934–
The mathematical and physical modeling of primary metals processing operations.

"A Wiley-Interscience publication."
Includes bibliographies and index.

1. Ore-dressing—Mathematical models.
2. Ore-dressing—Models. 3. Metallurgy—Mathematical models. 4. Metallurgy—Models.
I. Evans, James W., 1943– II. Brimacombe, J. K., 1943–
III. Title.

TN500.S96 1987 622'.34 87-8232
ISBN 0-471-83720-2

Printed in the United States of America

10 9 8 7 6 5 4 3 2 1

PREFACE

In recent years, mathematical and physical modeling has been accepted as one of the standard tools in new process development and in process optimization studies. While the modeling literature is growing by leaps and bounds, as yet there does not exist a comprehensive treatment of modeling issues.

Our aim in writing this volume has been to provide a synthesis of our collective experience in the modeling field, which has involved both mathematical and physical modeling, and a more limited recourse to pilot plant studies. We feel quite strongly that these three should be regarded as complementary approaches to the study of process problems.

Wherever possible, we sought to illustrate the fundamental principles through concrete applications; indeed, the second half of the book is devoted to a quite detailed discussion of several specific processing systems.

In carrying out this work, the authors are indebted to students past and present and to colleagues from the modeling profession for help with the work and for inspiration. No such list can ever be complete, but we would like to acknowledge Dr. M. Choudhary, Dr. C. W. Chang, Professor H. Y. Sohn, Professor R. Moreau, Professor Y. Fautrelle, Professor C. Vives, Professor N. El-Kaddah, Mr. J. W. Hlinka, Prof. I. Samarasekara, Dr. P. Barr, Dr. J. Richards, and Mr. J. J. McKay for particularly helpful input.

Modeling is a very dynamic field. Major growth is anticipated for the near future, spurred on by the cost-effectiveness of this approach, particularly in light of the much more powerful computers and software packages that are becoming available. For this reason, future applications will become much more sophisticated than the examples we have been able to cite. The basic principles, however, should remain the same.

J. SZEKELY
J. W. EVANS
J. K. BRIMACOMBE

Cambridge, Massachusetts
Berkeley, California
Vancouver, British Columbia

May, 1987

CONTENTS

1 INTRODUCTION

The purpose of this chapter is to introduce the mathematical and physical modeling techniques that have been developed to represent primary metals processing operations. These mathematical and physical modeling techniques have been proven effective both for diagnostic purposes—that is, troubleshooting, process improvement, and process optimization—and in the development of new technologies. The application of these modeling techniques to the solution of specific problems in metals processing is well documented in the literature; in contrast, no uniform treatment exists at present that clearly defines the fundamental principles used in a metallurgical context. Our aim is to emphasize the principles common to all the applications and then to illustrate the application of these principles through the use of specific examples.

The text is divided into five main sections: 1. an overview of mathematical and physical modeling; 2. the principles of mathematical modeling; 3. the principles of physical modeling; 4. a review of measurement techniques; and 5. illustrative industrial examples, including the modeling of continuous casting, aluminum electrolysis cells, and fluid–solid reactors.

1.1. AN OVERVIEW OF MATHEMATICAL AND PHYSICAL MODELING

The 1980s heralded profound technological change in the metals and materials field, driven by strong competitive pressures. These pressures are, in essence, twofold: (1) competition among individual national economies of developed, developing and Third World countries regarding the supply of a given commodity, such as steel, aluminum, or copper; and (2) competition between different materials—such as steel versus aluminum, or aluminum versus plastics or composites—within a given national economy. In either case it is recognized that, in order to remain competitive, it is essential to continually update and optimize existing metals processing operations and, perhaps more important, to continually develop new products and processes.

The primary metals industry is faced with particularly difficult problems in most developed countries, attributable to the very nature of these operations, namely, that they are capital intensive, energy intensive, and labor intensive, and produce a relatively small added value. This point is illustrated in Table 1.1, which gives typical values for the capital cost, operating cost, desired selling price, and actual selling price for some typical metallic materials.

Table 1.1 should be regarded as a rough guideline only, because of the marked fluctuations in current metal prices. Nevertheless, the table is highly instructive because it shows that in virtually all cases the desired selling price—that is, the price that would have to be met in order to recoup the capital investment and turn a reasonable profit—is much higher than the actual selling price. This means that, employing current technology, one could not justify the construction of new metals production plants in most developing countries. This contention is borne out in practice.

It follows that major changes are needed in the way that we produce and process metallic materials; indeed, a massive research and development effort aimed at both improving existing operations and, more important, developing new product and processing concepts would be crucial for the continued well-being of these industries and for the society as a whole, which is being served by these industrial enterprises. High specific production rate, automation, and robotization should be key components of such efforts.

Clearly, the time that one can allow for these developments will have to be quite short, so that the traditional evolutionary techniques will have to give way to alternative, speedier, and more cost-effective approaches. The mathematical models and, to a lesser extent, the physical modeling approaches to be discussed in this text are thought to have a great potential in aiding the accomplishment of this needed technological change.

Furthermore, as illustrated in Figure 1.1, in addition to the push by the need for technological change, there also exists a strong pull by the more cost-effective availability of computational tools, which include not only the hardware but also the various software packages.

As illustrated in Figure 1.2 the cost of performing a single calculation using a digital computer has markedly dropped during the past decade, and we can expect this trend to continue. It follows that, with the passage of time, mathematical modeling will become a progressively more cost-effective indeed indispensable, component of any process development effort.

While the principal illustrations in the text concern primary metals processing operations, the same concepts should be readily applicable to a broad range of materials processing systems.

TABLE 1.1 Capital Cost of Metals Producing Facilities, with Operating Cost, Required Selling Price, and Actual Selling Price[a]

Metal	Capital Cost $/Annual Ton Capacity	Operating Cost ($/Ton) (Excluding Depreciation)	Desired Selling Price ($/Ton) (Including Return on Investment and Taxes)	Actual Selling Price ($/Ton)
Aluminum	3,500	1,170	1,872	1,760
Copper	10,500	1,690	3,790	1,870
Steel (sheet)	1,500	400	720	450
Nickel	18,000	3,750	7,350	6,160
Lead	2,100	343	763	840
Zinc	2,300	600	1,060	840

SOURCE: Agarwal and Sinek.[1]

[a]For the very high interest rates prevailing in the U.S. at the time of writing (i.e., a prime rate of 10.5%) the capital charges would be much higher, resulting in rather higher values than cited.

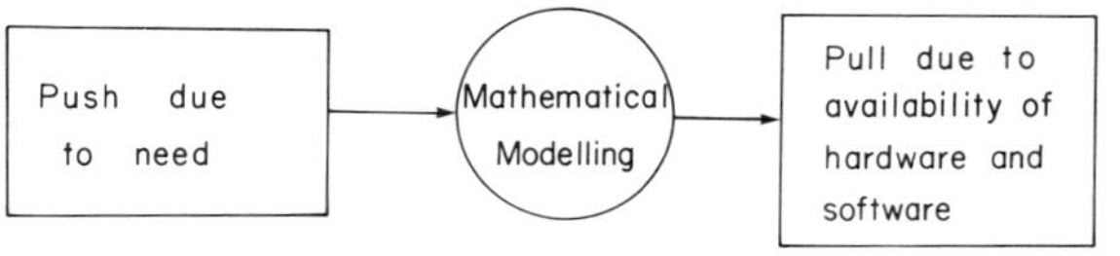

FIG. 1.1. The technological push and the industrial pull propelling mathematical modeling efforts.

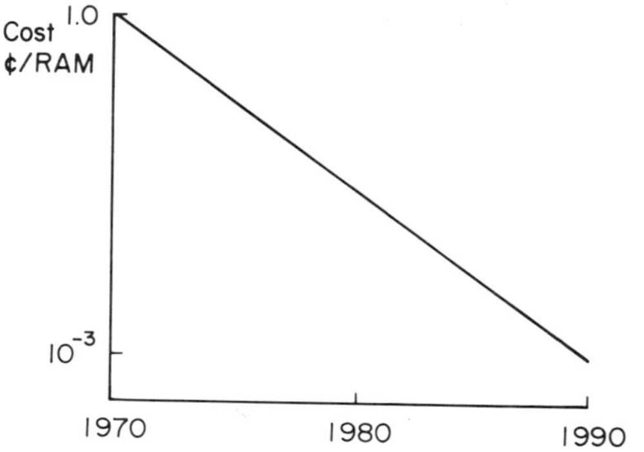

(*a*)

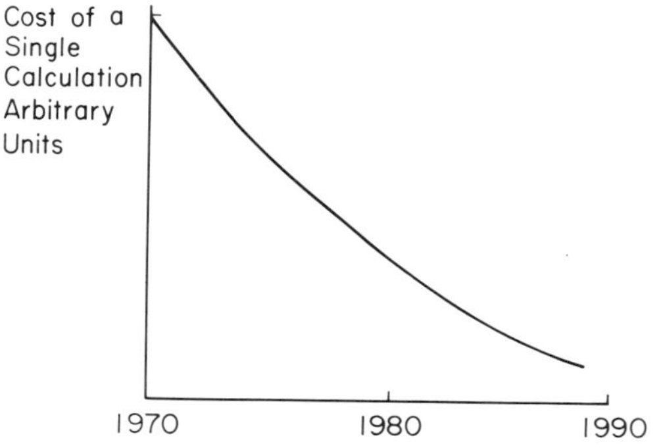

(*b*)

FIG. 1.2. (*a*) The evolution of the cost per bit of RAM (Random access memory) in cents. (*b*) The evolution of the cost of a given calculation, arbitrary scale.

1.2. MATHEMATICAL MODELS

In general, a mathematical model consists of a set of algebraic or differential equations that quantitatively represent a process or some aspect of a process. A mathematical model may be a relationship that defines the time required to empty a ladle holding molten steel or the time required for an ingot to solidify. In other contexts, a mathematical model may predict the total oxygen that has to be blown to complete a heat in a basic oxygen furnace (BOF) or the oil injection rate that is required to replace a certain amount of coke in an iron blast furnace.

1.2.1. Classification of Mathematical Models

Mathematical models may be classified into three main groups: (1) fundamentally based theoretical models, also termed mechanistic models; (2) semiempirical models; (3) and input–output type, "black-box" models.

Fundamentally based theoretical models are derived from basic physical laws, with a minimum amount of fudging, or empirical adjustment. For some certain simple physical situations it is possible to construct a fundamentally based mathematical model. In the majority of cases this will consist of partial differential equations that, together with their appropriate boundary conditions, have to be solved numerically.

Semiempirical models are based on physical laws but incorporate a certain amount of empiricism, either because some basic data are lacking or because of the complexity of the modeling equations, which would be difficult to solve. The vast majority of the mathematical models used in practice fall into this category.

Input–output type ("black-box") models are not based on physical laws but, rather, represent a totally empirical relationship between the key process variables. The relatively simple black-box type models are helpful in process control.

The difference between these three types of models is illustrated in the following example. Let us consider a sherical metal particle of radius R that is initially at a temperature T_i and that from time 0 loses heat by radiation to the environment, which is at a temperature T_e (both on the absolute scale), as sketched in Figure 1.3. The time-dependent temperature of this spherical particle is given by the following differential equation:

$$\frac{\partial T}{\partial t} = \alpha\left(\frac{2}{r}\frac{\partial T}{\partial r} + \frac{\partial^2 T}{\partial r^2}\right) \qquad (1)$$

with

$$T = T_i \qquad \text{at} \quad t = 0 \qquad (2)$$

$$\frac{\partial T}{\partial r} = 0, \qquad r = 0 \qquad (3)$$

$$-k\frac{\partial T}{\partial r} = \sigma\varepsilon(T^4 - T_e^4) \qquad \text{at} \quad r = R \qquad (4)$$

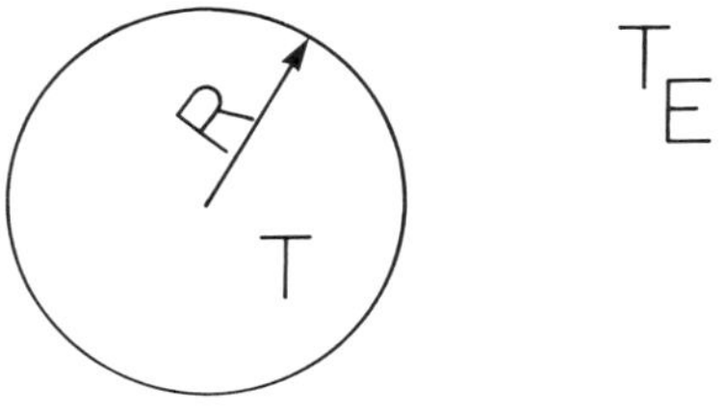

FIG. 1.3. Radiative heat transfer from a sphere.

where T is the absolute temperature,
t is time,
$\alpha = k/\rho C_p$ is the thermal diffusivity,
k is the thermal conductivity,
σ is the Stefan–Boltzmann constant, and
ε is the emissivity of the surface.

This system of equations may be solved numerically.[2]

Let us now consider the case, sketched in Figure 1.4, where spherical particles contained in a fluidized bed undergo a radiative heat exchange with a wall. For argument's sake, let us assign a temperature T_e to the wall and an initial temperature T_i to the particles. The resultant differential equation

$$\frac{\partial T}{\partial t} = \alpha \nabla^2 T \tag{5}$$

with $q_r = f(x, y, z, t)$ at the surface, would be quite cumbersome to solve. Furthermore, there is some question, as to how accurately one would be able to define the relationship expressed in Eq. (5).

For this reason one may wish to simplify the statement of the problem by

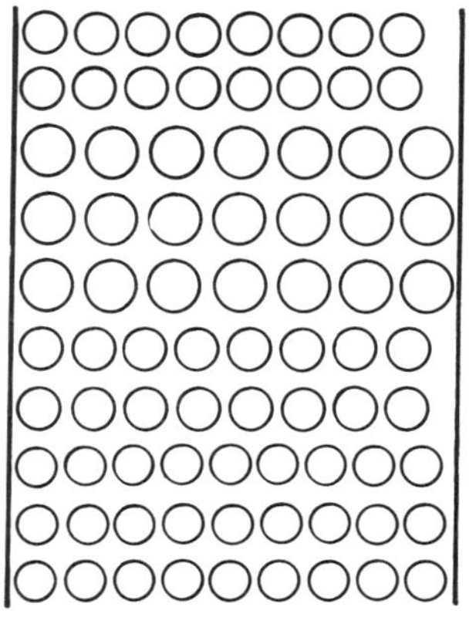

FIG. 1.4. Sketch of the fluidized bed system.

writing

$$\frac{\partial T}{\partial t} = \alpha\left(\frac{2}{r}\frac{\partial T}{\partial r} + \frac{\partial^2 T}{\partial r^2}\right) \tag{6}$$

with

$$T = T_i \qquad \text{at} \quad t = 0 \tag{7}$$

$$\frac{\partial T}{\partial r} = 0 \qquad \text{at} \quad r = 0 \tag{8}$$

and

$$-k\frac{\partial T}{\partial r} = F\sigma\varepsilon(T^4 - T_e^4) \qquad \text{at} \quad r = R \tag{9}$$

Where F is some sort of a "view factor" that incorporates the radiative interaction between the sphere and its surroundings.

The modeling of the system through Eqs. (6)–(9) is not quite rigorous because we glossed over the fact that the radiative interchange between the particle and its surroundings is spatially nonuniform. In other words, the isotherms would not be symmetrical about the center.

Yet another simplification may be possible if we postulate that the temperature field within the spherical particle would be uniform. This "lumped parameter" approach would yield the following, much simpler representation of the heat flow problem:

$$-4R^3\pi\rho C_p\frac{dT}{dt} = 4\pi R^2\varepsilon\sigma F(T^4 - T_e^4) \tag{10}$$

with

$$T = T_i \qquad \text{at} \quad t = 0 \tag{11}$$

In essence the two latter equations may be regarded as semiempirical representations of the problem. The key physical features of the system are incorporated into the model (i.e., the radiative heat exchange), but certain simplifications have been introduced (i.e., by neglecting the spatial nonuniformity of the flux). Most of the mathematical models used to represent metallurgical systems tend to fall in this category. The degree of simplification that is desirable depends on the nature of the problem, on the time available for generating a solution, and on the computational skills of the investigator.

The modeling of the system, depicted in Figure 1.4, through the use of the black-box approach would be to postulate that

$$T = f(T_e, T_i, [k, C_p, \rho, \varepsilon], t) \tag{12}$$

and then to find some empirical way to express the relationship, perhaps initially postulating a linear relationship between temperature and time (which would be certainly wrong) and then adjusting this relationship, for example, by the use of series expansion. The use of statistical analysis of data, time series techniques, and so on could be useful in the development of models of this type.[3,4]

In summary, a mathematical model represents a physical process or some aspect of a physical process in terms of a set of algebraic or differential equations. The main function of a mathematical model is to provide a working relationship among key process parameters, such as the time required for an ingot to solidify, the time required to empty a ladle, and the blast pressure required to maintain a given hot blast flow rate.

The principal usefulness of mathematical models lies in the following capabilities. For existing processes,

1. The model can improve our general understanding of the behavior of a process and the inter-relationships between key process variables.
2. The model can perform paper or computer studies exploring the effect of changing process variables and thus ultimately provide a means for optimizing the process.
3. The model can aid in the evaluation and the planning of plant trials.

For processes under development,

1. The model can help in the assessment of the feasibility of the process.
2. The model can help in the planning of laboratory scale experiments.
3. The model can help in the evaluation of pilot scale trials and scale-up.

1.3. PHYSICAL MODELS

If we wish to construct a mathematical model based on fundamentals, a necessary precondition is a good physical appreciation of the key process variables and the laws that govern the interactions among them. The model is the expression of these relationships in mathematical terms. In many cases these laws are known, as in the simple radiation heat transfer problem discussed in Section 1.2.1.

However, in many instances we do not have an adequate physical basis for the construction of a rigorous or even semirigorous mathematical model. In other cases the system of equations that would result from a rigorous model would be too complex to solve. Under such conditions we have another very useful tool at our disposal: the physical model.

In a physical model of a process, or of some particular aspect of a process, we seek to represent or model the system by changing the materials that are handled and also frequently the scale of the operation. *The main objective of modeling is to achieve a realistic representation of a system using materials and equipment with which measurements may be made conveniently and in a cost-effective manner.* The key phrases in this statement are *realistic representation* and *convenient measurement*, concepts that reflect the necessity of making compromises in certain instances.

Typical examples of physical modeling include the modeling of argon-stirred ladles containing molten steel and using a plexiglas vessel of, say, one-half the size of the industrial-scale unit containing water agitated by an airstream. Other examples are the representation of the velocity field in an induction furnace that uses mercury as the working fluid at a reduced scale and the modeling of the jet impact zone in a basic oxygen furnace employing the air–water system. In these examples, molten steel or hot combustion gases are represented by water at room temperature; thus, measurements can be made more readily. Furthermore, the reduction in linear scale allows measurements to be made in a cost-effective manner.

Physical model development may be classified in three categories:

1. A physical model may be constructed, following certain strict rules (see Chapter 3). Then the quantitative measurements made on the model may be translated directly, through appropriate scaling, to describe the behavior of real systems.
2. Alternatively, measurements in a corresponding physical system may be used to verify a mathematical model, and then the mathematical model may be used to predict the behavior of the real system.
3. One may also conduct a few ad hoc experiments, using a physical model, to obtain some insight into the behavior of the system.

All three approaches have a definite role in research and development studies. For example, properly scaled models have been used to deduce flue gas velocity fields in furnaces, and the velocity fields and mixing characteristics generated in ladles by natural convection. In the second category, elegant measurements of the rising velocities of bubbles in molten mercury and molten silver, conducted by Richardson and his co-workers,[5] have lead to the establishment of relationships with general validity. In the third category, tracer studies in real systems and in models of molten metals processing equipment have been invaluable in identifying some key features of the process, such as the existence of short-circuiting or bypass phenomena in copper reverberatory furnaces.

The choice of the approach depends on the nature of the system, on the nature of the problem, and on the time available.

1.4. PILOT PLANTS

Pilot plants constitute the third category in the classification of physical and mathematical modeling of metallurgical processes. A pilot plant is an exact replica of a full-scale installation except that it operates on a smaller scale, usually $\frac{1}{20}$–$\frac{1}{5}$ of the full-scale unit. Pilot plants are primarily used in new process development and usually represent the final stage, before the construction of a demonstration-scale, or commercial-scale unit. The objectives of pilot plants may be classified into the following three categories.

First, pilot plants are useful in new process development, since laboratory-scale measurements and physical and mathematical modeling studies usually do not enable one to design a full-scale plant with confidence. Thus, the construction of a processing unit with all the features of a full-scale installation, except size, is a necessary stage in process development. A pilot plant in this context should be used to provide the data for the design of the full-scale installation or a larger pilot unit. Frequently the pilot plant data will influence the decision as to whether to proceed with the project.

Second, pilot plants may be used in the study of existing processes. Often it is impractical to carry out large-scale plant trials on fully operating commercial processes because such trials would be costly and may interrupt production schedules. Pilot plants may be useful, therefore, in carrying out extensive tests with the view of optimizing the process.

Finally, pilot plant operations may provide product material suitable for a market survey. For example, a pilot plant can be used to provide form coke for blast furnace tests.

1.5. THE RELATIONSHIP BETWEEN MATHEMATICAL MODELS, PHYSICAL MODELS, AND PILOT PLANTS

The use of mathematical models, physical models, and pilot plants has been an indispensable component of virtually all metallurgical process development during the past two decades. Indeed, the progressively increasing use of mathematical and physical modeling has greatly improved the traditional route of proceeding from pilot plants to full-scale operations.

The development of and subsequent improvements in oxygen steel making, the bottom-blown oxygen converter operation (Q-BOP), continuous casting, ladle metallurgy, the new ALCOA smelting process for aluminum, and the Noranda process for continuous copper smelting have all been facilitated by the use of mathematical and physical modeling. While process development and process improvements have been effected in the past without recourse to mathematical or physical modeling, the more widespread application of these techniques has made research and development efforts much more cost effective by minimizing the hit-or-miss aspects of such programs.

Let us now consider the proper place of mathematical and physical modeling and pilot plant operations within the general hierarchy of process development and process improvement. It is perhaps a truism to say that most engineering problems have a multitude of solutions. Thus, one should consider mathematical and physical modeling techniques, in conjunction with pilot plant operation and full-scale plant trials, to represent a set of tools from which one might select, depending on the nature of the problem. In many instances one may wish to combine mathematical modeling with physical modeling before a pilot plant is constructed and then to use more extensive mathematical modeling for the interpretation of the pilot plant results and their ultimate scale-up.

In some instances pilot plant work will point to additional problems that may have to be resolved through additional physical modeling work aimed at some specific aspect of the process.

Experimental measurements that may be needed in order to make mathematical modeling meaningful include physical and chemical information on the system, such as basic thermodynamic data, data on reaction kinetics, and information on the values of such properties as thermal conductivity. A properly constructed mathematical model can be invaluable at the very beginning of the investigation to put these matters into perspective and to pinpoint the specific information that must be developed.

The following chapters introduce the reader to the elements of the mathematical and physical modeling techniques that are available, and illustrate how these techniques may be combined for tackling real process problems.

REFERENCES

1. J. C. Agarwal and J. R. Sinek, in Y. A. Chung, W. H. Danver, and J. M. Cigan, Eds., *Energy Use and Conservation in the Metals Industry*, A.I.M.E, New York, 1975.
2. J. Szekely and N. J. Themelis, *Rate Phenomena in Process Metallurgy*, Wiley, New York, 1971.
3. W. H. Ray and J. Szekely, *Process Optimization*, Wiley, New York, 1973, Chap. 1.
4. G. E. P. Box and R. E. Draper, *Evolutionary Operation*, Wiley, New York, 1969.
5. F. D. Richardson, *The Physical Chemistry of Melts in Metullurgy*, Academic Press, New York, 1974.

2 PRINCIPLES OF MATHEMATICAL MODELING

In combination with careful experimental measurements, a mathematical model can be a powerful tool in the development of new or existing metallurgical processes. Depending on the objectives of a study, a mathematical model offers many advantages:

1. It can increase understanding of a process in fundamental terms.
2. It can be used off line to study the influence of different variables on the operation of a process at a significantly lower cost than on the full-scale process.
3. It can assist in the evaluation of results from in-plant trials, for example, the conversion of temperature measurements to heat fluxes in a continuous-casting mold.
4. In an experimental campaign inplant, a model can assist in making decisions on trial conditions. For example, a model of a rotary kiln process might indicate that bed temperature in a particular region of the kiln is critical to productivity and product quality. Thus, experiments might concentrate on bed temperature measurements in this area rather than over the entire length of the kiln, which would be a formidable task on a plant scale.
5. A model may be an invaluable aid in scale-up and design.
6. A model can be used for the purpose of process control and optimization.

However, for a mathematical model to meet project objectives in a cost-effective manner, certain rules or procedures should be followed. Otherwise, as frequently happens, the model may not represent the physical and/or chemical phenomena in a process adequately, and the modeling activity is thus reduced to something less than a paper exercise. The goal of this chapter is to describe the role of mathematical modeling in process development; to present the basic steps of model building, including a brief discussion of numerical methods, and to offer examples of mathematical models based on published metallurgical studies.

2.1. TYPES OF MATHEMATICAL MODELS

Generally speaking a mathematical model is an equation or set of equations, algebraic or differential, together with initial and/or boundary conditions, that represents the important chemical and/or physical phenomena in a process. Mathematical models can be classified broadly into three categories.

2.1.1. Fundamental or Mechanistic Models

Fundamental models are based on mechanisms involving transport phenomena and chemical reaction rates in a process. Chemical and physical laws are applied to obtain mathematical equations relating independent and dependent variables. Examples of the laws involved are conservation of mass, energy, and momentum; Fourier's law (for conduction heat transfer), and Fick's law (for molecular diffusion). Frequently, empirical constants are required in the model to characterize phenomena that are very complicated and defy rigorous mathematical treatment. Many examples come to mind, such as the characterization of constants in an equation giving the kinetics of the austenite-to-pearlite transformation in a steel or the use of a convective heat transfer coefficient to describe mathematically the rate of cooling at the surface of a steel wire rod passing through an air blast on a controlled cooling line. In mass and heat transfer problems particularly, the characterization of boundary conditions with empirical transport coefficients is necessary and, as shall be seen, often requires that an experimental program be conducted in conjunction with the mathematical modeling work to measure the coefficients. Again, an example can be taken from the modeling of continuous casting, in which predictions of the temperature distribution in a strand cannot be made accurately until the rate of heat extraction from the surface of the strand is measured in the mold and spray zones.

This chapter is concerned only with this type of model, since it is most widely employed throughout the metallurgical industries.

2.1.2. Empirical Models

Empirical models rely on the determination of mathematical equations linking dependent and independent variables using measurements of the variables in an operating plant. The fitting of relationships is usually based on statistical analysis; but fundamental knowledge of the process may influence the form of the equations. However, unlike the fundamental model, the empirical model relies much less (often hardly at all) on an understanding of the mechanisms involved in the process. This type of model has been used extensively by mineral engineers to describe such processes as flotation, crushing, and grinding.

The empirical model, of course, requires a large number of measurements over a range of conditions, which is easier to achieve, for example, in a flotation circuit than in a rotary kiln operating at 1300°C. Depending on the process involved, the empirical model is often simpler and less expensive than the fundamental model, particularly where mechanisms in a process are poorly understood. However, owing to the method of formulation—that is, the fitting of equations to data—the empirical models are generally applicable only over the range of operation in which measurements have been made. Moreover, the models usually are valid only for the process units on which they were determined. In contrast, mechanistic models are normally applicable over a wider range of conditions than are empirical models, provided that the underlying mechanisms hold and empirical constants such as heat transfer coefficients are valid for those conditions. Finally—and this is important—the mechanistic model provides a much deeper fundamental understanding of the process than does the empirical model; this understanding is a great asset to the process engineer or researcher grappling with difficult operating problems.

2.1.3. Population-Balance Models

The population-balance model has been used extensively in the mineral processing field in recent years. It has been developed to describe processes having distributed properties, such as particle size in a ball mill. Population balances are performed on the number (or mass or volume) of entities (particles) having a particular value of a given property. The balances contain terms representing the arrival of entities at that particular property value, for example, the creation of a particular particle size by breakage of larger particles. The models are therefore characterized by the consideration of a distributed property rather than of the lumped properties, such as mass or momentum, treated in the fundamental, or mechanistic, models, described in Section 2.1.1. One of the few examples of the application of population-balance models to a metallurgical problem is in the study of inclusion coalescence and removal in molten steels. For more information on population-balance models, the reader is referred to such texts as Himmelblau and Bischoff.[1]

As shall be seen, there are other classifications of mathematical models, such as steady- or unsteady-state and lumped- or distributed-parameter models. These and other model types are described in detail by Himmelblau and Bischoff.[1]

2.2. ROLE OF MATHEMATICAL MODELS IN PROCESS ANALYSIS AND DEVELOPMENT

As already mentioned, a mathematical model can be a powerful tool in process development. However, it is not the only tool that the process engineer will need to employ in process analysis because a measurement program is normally necessary as well. Themelis[2] has described the types of measurements the process engineer may need to make in the study of a process:

1. Measurements of concentration, temperature, and fluid flow or mixing (tracer tests) in an existing process; techniques are described in Chapter 4.
2. Measurements on pilot plants that are designed according to four states of similarity with actual plants: geometric, mechanical, thermal, and chemical. Similarity criteria are discussed in Chapter 3.
3. Measurements on physical models that are designed to obey only one or two similarity criteria, for example, on a one-quarter–scale sectional model of a copper converter.

These types of measurements are not only complementary to mathematical modeling but are usually absolutely vital to determining empirical constants in the model, as related earlier, and are needed to check the adequacy of model predictions. Thus, an integrated approach to process analysis involves both mathematical modeling and measurements in plant or in the laboratory on a smaller scale. One example of this integrated approach is the mathematical modeling of the Stelco-Lurgi/Republic-National Lead direct-reduction process,[3] in which iron ore pellets are reduced with coal in a rotary kiln. The model required the measurement in the laboratory of empirical constants to characterize coal reactivity, iron ore reducibility, and the rate of coal devolatilization. In a 100 tons per day (tpd) pilot plant, measurements were made of bed temperature and waste gas analysis to check against model predictions for various reductant coals as part of the model validation. This aspect of model development has been discussed by Meadowcroft and Brimacombe.[4]

In many respects the mathematical model acts as a vital link between the research laboratory and the operating process. The model incorporates such parameters as transport coefficients, chemical rate constants, thermodynamic data, thermophysical properties, mechanical properties, and rheological properties, which are determined in the laboratory, and makes predictions of process behavior in plant. The SL/RN model mentioned above is but one example.

2.3. DEVELOPMENT OF A MATHEMATICAL MODEL

Whether a mathematical model is inherently complex or simple, its development should proceed through five stages: preparation, mathematical formulation, solution of the equations, validation of the model, and application of the model.

2.3.1. Preparation

In common with other techniques of process analysis, the development of a mathematical model should begin with a clear set of objectives. Is a model of a direct-chill casting machine for zinc required simply to predict the temperature distribution in a jumbo ingot with increased process knowledge as the objective? Or is the model required to provide a solution to a cracking problem in the jumbos? This is a very important state of model development; and in an industrial setting it requires that all parties involved—operating, quality control, research or industrial engineering, and so on—are consulted before proceeding. Input from these groups, coupled with a thorough discussion of the problem, is essential to ensure that it is well defined. Based on this discussion, it may be decided that mathematical modeling is not the best path to follow but that a physical modeling approach would be preferable. Or it may emerge that the problem was ill-defined to start with and that a different aspect of the process than originally envisaged needs to be mathematically modeled. The objectives that are finally agreed upon should be written down to ensure that the model development continues on track.

During this stage of problem formulation, a clear idea of the complexity of the mathematical model should emerge, and it should be borne in mind that the model should be made as simple as possible while maintaining a strong hold on reality. Does the model have to predict several dependent variables—such as temperature, concentration, velocity, and pressure—or is only one of these important? Does the model need to be three-dimensional, or can it be simplified to one or two dimensions without losing a hold on the reality of the process? Does the model need to be formulated to predict unsteady-state behavior, or is the assumption of steady-state valid? In addition, questions need to be asked concerning the property data, thermodynamic information, and transport coefficients required in the mathematical model. Are adequate data available to characterize the phenomena of interest, or can they be estimated with reasonable accuracy? What needs to be measured? Finally, the question of model validation must be considered. What measurements are necessary, and will they be performed in the laboratory or in plant? These questions need to be answered in order to estimate the time required to complete the modeling work, the cost of model development and related experimental work, and the likely success of the model in meeting the original objectives. These factors all bear on whether to proceed with a mathematical model.

2.3.2. Mathematical Formulation

In many respects, mathematical formulation is the easiest phase of model development, particularly if the preparation stage has been completed carefully. At this point, many decisions on the model should already have been made, such as those concerning steady or unsteady state, the coordinate system to be employed, the number of dimensions likely to be necessary, and the need to employ variable properties.

A. Assumptions

In further setting out a mathematical description of the process, it is usually necessary to make a number of assumptions. These may relate to the state of mixedness (backmixed or plug flow) of a fluid, the initial temperature of a solid, or the radiative properties of a surface (black or gray body). Assumptions of this kind are necessary to overcome gaps in our knowledge (e.g., the radiative properties of surfaces) or to make a difficult problem mathematically tractable; thus, the gases in the freeboard of a rotary kiln are normally assumed to move axially in plug flow because a mathematical description incorporating radial effects is simply too difficult. Depending on the complexity of the process and the objectives of the modeling study, the number of assumptions made may be very large. However, with each assumption, the hold of the model on reality is lossened; and, not in-frequently, a set of bad assumptions leads to the formulation of a model that bears no relationship to the real process and is incapable of achieving the stated objectives. Thus, this is a critical phase in model development, in which each assumption is examined critically relative to the process and, if possible, an estimate made of the effect of the assumption on the ability of the model to make reliable predictions. Once all assumptions have been made and evaluated, it is wise to reduce them to written form alongside the project objectives. This list of assumptions provides a constant reminder that the mathematical model is not reality but an attempt (sometimes poor) to describe reality in mathematical terms. This fact later becomes important again when model predictions are being examined and interpreted in light of process behavior.

B. The Balance

A mechanistic model must conform to the fundamental laws of conservation, and this is the basis of the mathematical formulation. The quantity to be conserved may be mass, heat, energy, or momentum, depending on the nature of the model. In order to apply conservation, a part of the system must be isolated, and thus a volume element must be defined. Four types of volume elements are possible, as shown in Figure 2.1 (for the Cartesian coordinate system). The choice of element depends on the number of independent space variables that are required to solve the problem. As shown in Figure 2.1, the use of an infinitesimal volume will lead to equations expressed in terms of three space variables, an infinitesimal area to two space variables, an infinitesimal section to one space variable, and an overall macrovolume will yield equations that contain no space variables. Of course, more information may be obtained from the three-dimensional case, but it is gained at the expense of increased computational cost and sometimes computational difficulties. Therefore, it is preferable to employ the minimum number of dimensions consistent with an adequate mathematical description of the process and to choose the volume element accordingly.

Once the volume element has been defined, a balance is performed on the quantity to be conserved (e.g., heat, mass, or momentum). The balance can be

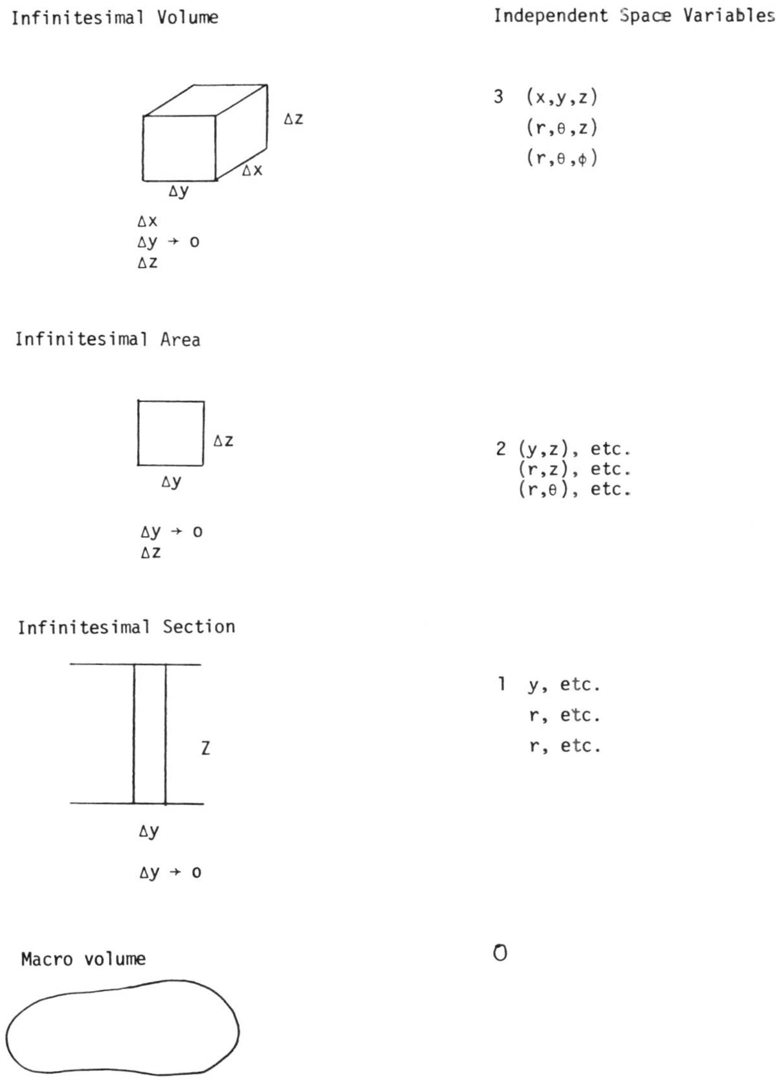

FIG. 2.1. Types of volume elements.

written generally as

$$\begin{pmatrix}\text{input through surfaces}\\ \text{of volume element}\end{pmatrix} - \begin{pmatrix}\text{output through surfaces}\\ \text{of volume element}\end{pmatrix} + \begin{pmatrix}\text{generation inside}\\ \text{volume element}\end{pmatrix} - \begin{pmatrix}\text{consumption inside}\\ \text{volume element}\end{pmatrix} = \begin{pmatrix}\text{net accumulation in}\\ \text{volume element}\end{pmatrix} \tag{1}$$

The first two terms in Eq. (1) include all important transport mechanisms by which the conserved quantity moves *into or out of* the volume element. These mechanisms are bulk motion (convection) and diffusion for heat, mass, and momentum; and in the case of heat transfer, a third mechanism, radiation, may also be operative. The mathematical description of each of these transport mechanisms is given for Cartesian coordinates in Table 2.1. Usually in the balance, the relevant flux terms from Table 2.1 are multiplied by the area of the normal face of the volume element to yield a rate. This will be shown in ensuing examples (Section 2.4). These terms give the space dependency of the dependent variable in question (e.g. temperature, molar concentration, and velocity).

The second two terms in Eq. (1) account for changes to the conserved quantity *within* the volume element. Thus, for instance, heat may be generated or consumed by a homogeneous chemical reaction occurring within the confines of the volume element. Phase transformations in solid steel are another example of a process that generates or consumes heat throughout the volume. Moles of a chemical species may be destroyed or generated owing to a homogeneous chemical reaction, but mass (in the nonrelativistic world of the process metallurgist) cannot be consumed or created. In the case of momentum transfer, the equivalents of consumption or generation terms are forces due to gravity and an electromagnetic field. Generation and consumption terms are normally expressed as rates per unit volume and in the balance are multiplied by the volume of the volume element, resulting in a rate.

The final term in the balance expresses the buildup or depletion of the conserved quantity within the volume element and gives the change of the dependent variable with time. If this term is zero, the process, by definition, is at steady state.

The balances yield the governing equations that relate the dependent

TABLE 2.1. Characterization of Input–Output Terms in the Balance (Cartesian Coordinates)

Heat flux	
Bulk motion	$u\rho C_p T$
Thermal diffusion	$-k\partial T/\partial x$
Radiation	$-4\sigma T^3 \partial T/\partial x$
Mass flux	
Bulk motion	$u\rho_i$ or uC_i
Molecular diffusion	$-D_{i,m}\partial\rho_i/\partial x$ or $-D_{i,m}\partial C_i/\partial x$
Momentum flux	
Bulk motion	$\rho u_x u_x$, $\rho u_x u_y$, $\rho u_x u_z$
Viscous diffusion[a]	$-\mu[2\partial u_x/\partial x - 2/3\nabla\cdot u]$

[a]For a list of the equations giving the diffusive component of momentum flux (i.e., the stress tensor) for a Newtonian fluid, the interested reader should consult Bird et al.[5]

variables—concentration from mass balances, temperature from heat balances, and velocity and pressure from momentum balances—to the independent variables: space and time. The equation may be a partial or an ordinary differential equation, depending on the number of independent variables; or the equation may be a simple algebraic equation if the dependent variable does not vary with space and time. The methods of solving differential equations are discussed briefly in Sections 2.3.3.C and 2.3.3.D.

C. Boundary and Initial Conditions

Mathematical formulation of a process is not complete without the specification of initial conditions (for unsteady-state problems) and boundary conditions, subject to which the governing equation is solved. A set of n boundary conditions is required for each nth-order derivative in the differential equation, and an initial condition is needed for each time derivative, which usually is first order. For example, in the case of two-dimensional, unsteady-state heat conduction in a slab in a reheating furnace for which the governing equation is

$$\frac{\partial}{\partial x}\left(k\frac{\partial T}{\partial x}\right) + \frac{\partial}{\partial y}\left(k\frac{\partial T}{\partial y}\right) = \rho C_p \frac{\partial T}{\partial t} \tag{2}$$

four boundary conditions (two for each second-order derivative) and one initial condition need to be specified.

The specification of a particular boundary condition depends on the local conditions at the boundary; in other words, the problem at hand will usually tell the process engineer which kind of boundary condition is most appropriate. There are three types of boundary conditions.

1. SPECIFICATION OF THE DEPENDENT VARIABLE AT THE BOUNDARY. A good example is the "no-slip" condition at a stationary solid boundary past which a fluid is flowing:

$$\begin{matrix} x = 0, & u_0 = 0 \\ \text{(at solid surface)} & \text{(fluid velocity is zero)} \end{matrix} \tag{3}$$

Another example is the specification of temperature at the surface of a slab:

$$x = 0, \qquad T = T_0 \qquad \text{or} \qquad T = T_0(t) \tag{4}$$

although in most heating and cooling problems this boundary condition is not likely to be as useful or realistic as other types (see Section 3 below).

2. EXPRESSION OF EQUILIBRIUM AT THE BOUNDARY BETWEEN TWO PHASES. For heat transfer problems in which there is no contact resistance at the boundary between two phases, the temperature of the phases immediately on either side of

the boundary must be equal (i.e., there is thermal equilibrium locally at the boundary). Thus, for phases 1 and 2 in contact

$$x = 0, \qquad T_{01} = T_{02} \tag{5}$$

Another case in which local equilibrium may be specified is that of mass transfer accompanied by an "instantaneous" chemical reaction at the boundary between two phases, such as slag and metal or metal and gas. This is a problem of considerable importance to the process engineer concerned with pyrometallurgical extraction and refining operations because the chemical reactions involved are mostly heterogeneous and, at elevated temperatures, are usually sufficiently fast to be considered instantaneous. Under these conditions, chemical equilibrium prevails locally at the boundary, so the interfacial concentrations of reactants and products are related by the equilibrium constant. For the general reaction

$$a\mathrm{A} + b\mathrm{B} = d\mathrm{D} + e\mathrm{E} \tag{6}$$

the equilibrium constant, written in terms of molar concentrations at the interface, C^*, is

$$\Omega = \frac{(C_{\mathrm{D}}^*)^d (C_{\mathrm{E}}^*)^e}{(C_{\mathrm{A}}^*)^a (C_{\mathrm{B}}^*)^b} \tag{7}$$

Turning to momentum transfer, one might specify that the velocities of two fluids on either side of a phase boundary are equal.

3. Continuity of Heat, Mass, or Momentum at a Boundary Surface. In heat flow problems involving conduction, it is common to express as a boundary condition the continuity between conduction internally (described by Fourier's law) at the boundary and external heat transfer from and to the boundary. In mathematical terms, this boundary condition is written as follows:

$$x = 0, \qquad -k\left.\frac{\partial T}{\partial x}\right|_{x=0} = \dot{q}_0 \tag{8}$$

where, depending on the problem, $\dot{q}_0$ may be a surface heat flux, which frequently must be characterized empirically. For example, in developing a mathematical model of continuous casting, the rate of heat extraction by the mold is determined as a function of axial position (or time) by inserting thermocouples into the copper wall of the mold under different casting conditions (lubricant type, steel composition, and casting speed). The heat flux can be substituted directly into Eq. (8), or, alternatively, it may be expressed in terms of a heat transfer coefficient multiplying a temperature driving force (e.g., temperature difference between steel surface and mold wall in a continuous-casting machine). The latter is particularly useful when correlations exist between the heat transfer coefficient and the flow conditions and fluid properties, as in the case of the air-cooling of wire rod on a controlled cooling line. These correlations are usually expressed in terms of dimensionless variables such as Nusselt number, Reynolds number, and Prandtl number (dimensionless numbers are discussed in detail in Chapter 3). In other instances the heat flux in Eq. (8) may be based on a fundamentally derived equation such as the Stefan–Boltzmann law for radiative heat exchange. Thus, the surface boundary condition for a slab losing heat to the environment predominantly by radiation would be

$$x = 0, \qquad -k \left.\frac{\partial T}{\partial x}\right|_{x=0} = \sigma\varepsilon(T_0^4 - T_a^4) \tag{9}$$

where σ is the Stefan–Boltzmann constant,
ε is the emissivity of the slab surface, and
T_a is the ambient temperature.

For interior boundaries, such as between two refractory layers (1 and 2) in a furnace wall, continuity of heat conduction on either side of the boundary can be expressed as

$$x = 0, \qquad -k_1 \left.\frac{\partial T_1}{\partial x}\right|_{x=0} = -k_2 \left.\frac{\partial T_2}{\partial x}\right|_{x=0} \tag{10}$$

Another example of this type of boundary condition in heat conduction models is the no-flux, or adiabatic, surface, for which Eq. (8) becomes

$$x = 0, \qquad -k \left.\frac{\partial T}{\partial x}\right|_{x=0} = 0 \tag{11}$$

This boundary condition is used frequently to express symmetry of heat conduction, as would be found at the axis of a rod being cooled uniformly or at the axial midplane of a rectangular ingot being heated uniformly. Under such conditions heat does not flow across the axis or midplane, and the temperature distribution is symmetrical about that position. If valid, the no-flux boundary condition is very useful in modeling because it permits a substantial reduction in the space over which the dependent variable needs to be predicted. For example, in a heat flow model of continuous billet casting, the application of two vertical planes of symmetry intersecting the midface of the rectangular (square) cross section reduces computation by a factor of 4.

With respect to mass transfer situations, analogous boundary conditions can be written. In addition, it is worth mentioning that when mass transfer is accompanied by a heterogeneous reaction, the fluxes of reactants and products to and from the boundary, in the absence of accumulation, are related by the stoichiometry of the reaction. Moreover, if the chemical reaction is intrinsically slow, so that local equilibrium does not prevail, the rate of the reaction at the boundary must be specified.

In momentum transfer models, boundary conditions of this type would be used at fluid–fluid boundaries where continuity of momentum flux would be specified. Moreover, in the case of gas–liquid boundaries, a no-flux condition can be specified (i.e., the gas exerts negligible drag on the liquid).

As can be seen, the characterization of boundary conditions is a relatively straightforward task, but the determination of fluxes, transport coefficients, and thermodynamic data (e.g., activity coefficients) when they are not available often poses a formidable problem. It cannot be overemphasized that, unless boundary conditions are well characterized, the model predictions may be erroneous and misleading. Thus, as stated earlier, an experimental program, conducted in the laboratory and/or in the plant, frequently must be included in the development of a mathematical model to determine boundary conditions.

Finally, the initial conditions for unsteady-state models must be specified based on a knowledge of the starting point of a process (e.g., the temperature distribution in a slab as it leaves a reheating furnace ready to be rolled).

As described earlier, the balance leads to the formulation of governing equations in the mathematical model. The explicit application of the procedure given can be followed in a straightforward manner for relatively simple problems. However, for more complicated situations, the alternative approach may be to use the appropriate transport equations as the starting point, which may be then regarded as the building blocks of the mechanistic mathematical model. Examples of both approaches are given in Section 2.4. However, at this stage it may be worthwhile to present a very brief review of the basic transport equations, which may be regarded as the tool kit of mathematical modelers.

D. Fluid Flow

Laminar fluid flow problems—these, incidentally, are quite rarely encountered in process metallurgy practice (with the exception of entrainment, the flotation of inclusions, lubrication, and the movement of viscous slags)—may be represented by writing down the equation of continuity, which is in essence a mass balance:

$$\nabla \cdot \boldsymbol{\rho} \mathbf{U} = 0 \tag{12}$$

and the equation of motion, often termed the Navier–Stokes equation, which is a differential momentum balance:

$$\underset{\substack{\text{(accumulation of}\\ \text{momentum)}}}{\frac{\partial(\rho \mathbf{U})}{\partial t}} = \underset{\substack{\text{(convective transport}\\ \text{of momentum)}}}{-\nabla \cdot \rho \mathbf{U}\mathbf{U}} + \underset{\substack{\text{(diffusive transport}\\ \text{of momentum)}}}{\nabla \mu \nabla \mathbf{U}} + \underset{\text{(body force)}}{\rho \mathbf{F}_b} - \underset{\substack{\text{(force due}\\ \text{to pressure)}}}{\nabla P} \tag{13}$$

where ρ is fluid density,

> $\mathbf{U}$ is the velocity vector, which has three components (u_x, u_y, and u_z),
> μ is the molecular viscosity,
> $\mathbf{F_b}$ is the body force vector, which has three components (F_x, F_y, and F_z), and may be due to gravity and/or electromagnetic forces, and
> P is pressure.

The equivalent expressions for Eqs. (12) and (13) in Cartesian, cylindrical, and spherical coordinates are given in Table 2.2.

TABLE 2.2. The Transport Equations

Equations of Motion in Rectangular Coordinates (x, y, z)

x component
$$\rho\left(\frac{\partial u_x}{\partial t} + u_x\frac{\partial u_x}{\partial x} + u_y\frac{\partial u_x}{\partial y} + u_z\frac{\partial u_x}{\partial z}\right) = -\frac{\partial p}{\partial x} - \left(\frac{\partial \tau_{xx}}{\partial x} + \frac{\partial \tau_{yx}}{\partial y} + \frac{\partial \tau_{zx}}{\partial z}\right) + \rho g_x$$

y component
$$\rho\left(\frac{\partial u_y}{\partial t} + u_x\frac{\partial u_y}{\partial x} + u_y\frac{\partial u_y}{\partial y} + u_z\frac{\partial u_y}{\partial z}\right) = -\frac{\partial p}{\partial y} - \left(\frac{\partial \tau_{xy}}{\partial x} + \frac{\partial \tau_{yy}}{\partial y} + \frac{\partial \tau_{zy}}{\partial z}\right) + \rho g_y$$

z component
$$\rho\left(\frac{\partial u_z}{\partial t} + u_x\frac{\partial u_z}{\partial x} + u_y\frac{\partial u_z}{\partial y} + u_z\frac{\partial u_z}{\partial z}\right) = -\frac{\partial p}{\partial z} - \left(\frac{\partial \tau_{xz}}{\partial x} + \frac{\partial \tau_{yz}}{\partial y} + \frac{\partial \tau_{zz}}{\partial z}\right) + \rho g_z$$

For a Newtonain fluid with constant ρ and μ:

x component
$$\rho\left(\frac{\partial u_x}{\partial t} + u_x\frac{\partial u_x}{\partial x} + u_y\frac{\partial u_x}{\partial y} + u_z\frac{\partial u_x}{\partial z}\right) = -\frac{\partial p}{\partial x} + \mu\left(\frac{\partial^2 u_x}{\partial x^2} + \frac{\partial^2 u_x}{\partial y^2} + \frac{\partial^2 u_x}{\partial z^2}\right) + \rho g_x$$

y component
$$\rho\left(\frac{\partial u_y}{\partial t} + u_x\frac{\partial u_y}{\partial x} + u_y\frac{\partial u_y}{\partial y} + u_z\frac{\partial u_y}{\partial z}\right) = -\frac{\partial p}{\partial y} + \mu\left(\frac{\partial^2 u_y}{\partial x^2} + \frac{\partial^2 u_y}{\partial y^2} + \frac{\partial^2 u_y}{\partial z^2}\right) + \rho g_y$$

(continued on next page)

z component $$\rho\left(\frac{\partial u_z}{\partial t} + u_x \frac{\partial u_z}{\partial x} + u_y \frac{\partial u_z}{\partial y} + u_z \frac{\partial u_z}{\partial z}\right) = -\frac{\partial p}{\partial z} + \mu\left(\frac{\partial^2 u_z}{\partial x^2} + \frac{\partial^2 u_z}{\partial y^2} + \frac{\partial^2 u_z}{\partial z^2}\right) + \rho g_z$$

Equations of Motion in Cylindrical Coordinates (r, θ, z)

In terms of τ:

r component $$\rho\left(\frac{\partial u_r}{\partial t} + u_r \frac{\partial u_r}{\partial r} + \frac{u_\theta}{r}\frac{\partial u_r}{\partial \theta} - \frac{u_\theta^2}{r} + u_z \frac{\partial u_r}{\partial z}\right) = -\frac{\partial p}{\partial r} - \left(\frac{1}{r}\frac{\partial}{\partial r}(r\tau_{rr}) + \frac{1}{r}\frac{\partial \tau_{r\theta}}{\partial \theta} - \frac{\tau_{\theta\theta}}{r} + \frac{\partial \tau_{rz}}{\partial z}\right) + \rho g_r$$

θ component $$\rho\left(\frac{\partial u_\theta}{\partial t} + u_r \frac{\partial u_\theta}{\partial r} + \frac{u_\theta}{r}\frac{\partial u_\theta}{\partial \theta} + \frac{u_r u_\theta}{r} + u_z \frac{\partial u_\theta}{\partial z}\right) = -\frac{1}{r}\frac{\partial p}{\partial \theta} - \left(\frac{1}{r^2}\frac{\partial}{\partial r}(r^2\tau_{r\theta}) + \frac{1}{r}\frac{\partial \tau_{\theta\theta}}{\partial \theta} + \frac{\partial \tau_{\theta z}}{\partial z}\right) + \rho g_\theta$$

z component $$\rho\left(\frac{\partial u_z}{\partial t} + u_r \frac{\partial u_z}{\partial r} + \frac{u_\theta}{r}\frac{\partial u_z}{\partial \theta} + u_z \frac{\partial u_z}{\partial z}\right) = -\frac{\partial p}{\partial z} - \left(\frac{1}{r}\frac{\partial}{\partial r}(r\tau_{rz}) + \frac{1}{r}\frac{\partial \tau_{\theta z}}{\partial \theta} + \frac{\partial \tau_{zz}}{\partial z}\right) + \rho g_z$$

For a Newtonian fluid with constant ρ and μ:

r component $$\rho\left(\frac{\partial u_r}{\partial t} + u_r \frac{\partial u_r}{\partial r} + \frac{u_\theta}{r}\frac{\partial u_r}{\partial \theta} - \frac{u_\theta^2}{r} + u_z \frac{\partial u_r}{\partial z}\right) = -\frac{\partial p}{\partial r} + \mu\left[\frac{\partial}{\partial r}\left(\frac{1}{r}\frac{\partial}{\partial r}(ru_r)\right) + \frac{1}{r^2}\frac{\partial^2 u_r}{\partial \theta^2} - \frac{2}{r^2}\frac{\partial u_\theta}{\partial \theta} + \frac{\partial^2 u_r}{\partial z^2}\right] + \rho g_r$$

θ component $$\rho\left(\frac{\partial u_\theta}{\partial t} + u_r \frac{\partial u_\theta}{\partial r} + \frac{u_\theta}{r}\frac{\partial u_\theta}{\partial \theta} + \frac{u_r u_\theta}{r} + u_z \frac{\partial u_\theta}{\partial z}\right) = -\frac{1}{r}\frac{\partial p}{\partial \theta} + \mu\left[\frac{\partial}{\partial r}\left(\frac{1}{r}\frac{\partial}{\partial r}(ru_\theta)\right) + \frac{1}{r^2}\frac{\partial^2 u_\theta}{\partial \theta^2} + \frac{2}{r^2}\frac{\partial u_r}{\partial \theta} + \frac{\partial^2 u_\theta}{\partial z^2}\right] + \rho g_\theta$$

z component $$\rho\left(\frac{\partial u_z}{\partial t} + u_r \frac{\partial u_z}{\partial r} + \frac{u_\theta}{r}\frac{\partial u_z}{\partial \theta} + u_z \frac{\partial u_z}{\partial z}\right) = -\frac{\partial p}{\partial z} + \mu\left[\frac{1}{r}\frac{\partial}{\partial r}\left(r\frac{\partial u_z}{\partial r}\right) + \frac{1}{r^2}\frac{\partial^2 u_z}{\partial \theta^2} + \frac{\partial^2 u_z}{\partial z^2}\right] + \rho g_z$$

Equations of Motion in Spherical Coordinates (r, θ, ϕ)

In terms of τ:

r component
$$\rho\left(\frac{\partial u_r}{\partial t} + u_r\frac{\partial u_r}{\partial r} + \frac{u_\theta}{r}\frac{\partial u_r}{\partial \theta} + \frac{u_\phi}{r\sin\theta}\frac{\partial u_r}{\partial \phi} - \frac{u_\theta^2 + u_\phi^2}{r}\right)$$
$$= -\frac{\partial p}{\partial r} - \left(\frac{1}{r^2}\frac{\partial}{\partial r}(r^2\tau_{rr}) + \frac{1}{r\sin\theta}\frac{\partial}{\partial\theta}(\tau_{r\theta}\sin\theta)\right.$$
$$\left.+ \frac{1}{r\sin\theta}\frac{\partial\tau_{r\phi}}{\partial\phi} - \frac{\tau_{\theta\theta}+\tau_{\phi\phi}}{r}\right) + \rho g_r$$

θ component
$$\rho\left(\frac{\partial u_\theta}{\partial t} + u_r\frac{\partial u_\theta}{\partial r} + \frac{u_\theta}{r}\frac{\partial u_\theta}{\partial \theta} + \frac{u_\phi}{r\sin\theta}\frac{\partial u_\theta}{\partial \phi} + \frac{u_r u_\theta}{r} - \frac{u_\phi^2\cot\theta}{r}\right)$$
$$= -\frac{1}{r}\frac{\partial p}{\partial \theta} - \left(\frac{1}{r^2}\frac{\partial}{\partial r}(r^2\tau_{r\theta}) + \frac{1}{r\sin\theta}\frac{\partial}{\partial\theta}(\tau_{\theta\theta}\sin\theta)\right.$$
$$\left.+ \frac{1}{r\sin\theta}\frac{\partial\tau_{\theta\phi}}{\partial\phi} + \frac{\tau_{r\theta}}{r} - \frac{\cot\theta}{r}\tau_{\phi\phi}\right) - \rho g_\theta$$

ϕ component
$$\rho\left(\frac{\partial u_\phi}{\partial t} + u_r\frac{\partial u_\phi}{\partial r} + \frac{u_\theta}{r}\frac{\partial u_\phi}{\partial \theta} + \frac{u_\phi}{r\sin\theta}\frac{\partial u_\phi}{\partial \phi} + \frac{u_\phi u_r}{r} + \frac{u_\theta u_\phi}{r}\cot\theta\right)$$
$$= -\frac{1}{r\sin\theta}\frac{\partial p}{\partial\phi} - \left(\frac{1}{r^2}\frac{\partial}{\partial r}(r^2\tau_{r\phi}) + \frac{1}{r}\frac{\partial\tau_{\theta\phi}}{\partial\theta} + \frac{1}{r\sin\theta}\frac{\partial\tau_{\phi\phi}}{\partial\phi}\right.$$
$$\left.+ \frac{\tau_{r\phi}}{r} + \frac{2\cot\theta}{r}\tau_{\theta\phi}\right) + \rho g_\phi$$

In terms of velocity gradients for a Newtonian fluid with constant, ρ and μ^a;

r component
$$\rho\left(\frac{\partial u_r}{\partial t} + u_r\frac{\partial u_r}{\partial r} + \frac{u_\theta}{r}\frac{\partial u_r}{\partial \theta} + \frac{u_\phi}{r\sin\theta}\frac{\partial u_r}{\partial \phi} - \frac{u_\theta^2 + u_\phi^2}{r}\right)$$
$$= -\frac{\partial p}{\partial r} + \mu\left(\nabla^2 u_r - \frac{2}{r^2}u_r - \frac{2}{r^2}\frac{\partial u_\theta}{\partial\theta} - \frac{2}{r^2}u_\theta\cot\theta\right.$$
$$\left.- \frac{2}{r^2\sin\theta}\frac{\partial u_\phi}{\partial\phi}\right) + \rho g_r$$

θ component
$$\rho\left(\frac{\partial u_\theta}{\partial t} + u_r\frac{\partial u_\theta}{\partial r} + \frac{u_\theta}{r}\frac{\partial u_\theta}{\partial \theta} + \frac{u_\phi}{r\sin\theta}\frac{\partial u_\theta}{\partial \phi} + \frac{u_r u_\theta}{r} - \frac{u_\phi^2\cot\theta}{r}\right)$$
$$= -\frac{1}{r}\frac{\partial p}{\partial\theta} + \mu\left(\nabla^2 u_\theta + \frac{2}{r^2}\frac{\partial u_r}{\partial\theta} - \frac{u_\theta}{r^2\sin^2\theta} - \frac{2\cos\theta}{r^2\sin^2\theta}\frac{\partial u_\phi}{\partial\phi}\right) + \rho g_\theta$$

(continued on next page)

ϕ component

$$\rho\left(\frac{\partial u_\phi}{\partial t} + u_r\frac{\partial u_\phi}{\partial r} + \frac{u_\theta}{r}\frac{\partial u_\phi}{\partial \theta} + \frac{u_\phi}{r\sin\theta}\frac{\partial u_\phi}{\partial \phi} + \frac{u_\phi u_r}{r} + \frac{u_\theta u_\phi}{r}\cot\theta\right)$$
$$= -\frac{1}{r\sin\theta}\frac{\partial p}{\partial \phi} + \mu\left(\nabla^2 u_\phi - \frac{u_\phi}{r^2\sin^2\theta} + \frac{2}{r^2\sin\theta}\frac{\partial u_r}{\partial \phi} + \frac{2\cos\theta}{r^2\sin^2\theta}\frac{\partial u_\theta}{\partial \phi}\right) + \rho g_\phi$$

Equations of Heat Conduction in the Absence of Fluid Flow (Constant k)

Cartesian coordinates

$$\alpha\left(\frac{\partial^2 T}{\partial x^2} + \frac{\partial^2 T}{\partial y^2} + \frac{\partial^2 T}{\partial z^2}\right) + \frac{\dot q(x, y, z)}{\rho C_p} = \frac{\partial T}{\partial t}$$

Cylindrical coordinates

$$\alpha\left(\frac{\partial^2 T}{\partial r^2} + \frac{1}{r}\frac{\partial T}{\partial r} + \frac{1}{r^2}\frac{\partial^2 T}{\partial \theta^2} + \frac{\partial^2 T}{\partial z^2}\right) + \frac{\dot q(r, \theta, z)}{\rho C_p} = \frac{\partial T}{\partial t}$$

Spherical coordinates

$$\frac{\alpha}{r^2}\left[\frac{\partial}{\partial r}\left(r^2\frac{\partial T}{\partial r}\right) + \frac{1}{\sin\theta}\frac{\partial}{\partial \theta}\left(\sin\theta\frac{\partial T}{\partial \theta}\right) + \frac{1}{\sin^2\theta}\frac{\partial^2 T}{\partial \phi^2}\right] + \frac{\dot q(r, \theta, \phi)}{\rho C_p} = \frac{\partial T}{\partial t}$$

Equations of Heat Conduction in the Presence of Fluid Flow (Constant k, Constant P, Negligible Heat Generation due to Viscous Forces)

Cartesian coordinates

$$\alpha\left(\frac{\partial^2 T}{\partial x^2} + \frac{\partial^2 T}{\partial y^2} + \frac{\partial^2 T}{\partial x^2}\right) - \left(u_x\frac{\partial T}{\partial x} + u_y\frac{\partial T}{\partial y} + u_z\frac{\partial T}{\partial z}\right) + \frac{\dot q(x, y, z)}{\rho C_p} = \frac{\partial T}{\partial t}$$

Cylindrical coordinates

$$\alpha\left[\frac{1}{r}\frac{\partial}{\partial r}\left(r\frac{\partial T}{\partial r}\right) + \frac{1}{r^2}\frac{\partial^2 T}{\partial \theta^2} + \frac{\partial^2 T}{\partial z^2}\right] - \left(u_r\frac{\partial T}{\partial r} + \frac{u_\theta}{r}\frac{\partial T}{\partial \theta} + u_z\frac{\partial T}{\partial z}\right) + \frac{\dot q(r, \theta, z)}{\rho C_p} = \frac{\partial T}{\partial t}$$

Spherical coordinates

$$\frac{\alpha}{r^2}\left[\frac{\partial}{\partial r}\left(r^2\frac{\partial T}{\partial r}\right) + \frac{1}{\sin\theta}\frac{\partial}{\partial \theta}\left(\sin\theta\frac{\partial T}{\partial \theta}\right) + \frac{1}{\sin^2\theta}\frac{\partial^2 T}{\partial \phi^2}\right]$$
$$-\left(u_r\frac{\partial T}{\partial r} + \frac{u_\theta}{r}\frac{\partial T}{\partial \theta} + \frac{u_\phi}{r\sin\theta}\frac{\partial T}{\partial \phi}\right) + \frac{\dot q(r, \theta, \phi)}{\rho C_p} = \frac{\partial T}{\partial t}$$

Equations of Diffusion in the Absence of Fluid Flow (Constant Diffusivity, Low Concentration of Diffusing Species)

Cartesian coordinates

$$D_{A-B}\left(\frac{\partial^2 X_A}{\partial x^2} + \frac{\partial^2 X_A}{\partial y^2} + \frac{\partial^2 X_A}{\partial z^2}\right) + \frac{\dot r_A(x, y, z)}{C_T} = \frac{\partial X_A}{\partial t}$$

Cylindrical coordinates

$$D_{\mathrm{A-B}}\left(\frac{\partial^2 X_{\mathrm{A}}}{\partial r^2}+\frac{1}{r}\frac{\partial X_{\mathrm{A}}}{\partial r}+\frac{1}{r^2}\frac{\partial^2 X_A}{\partial \theta^2}+\frac{\partial^2 X_A}{\partial z^2}\right)+\frac{\dot{r}_{\mathrm{A}}(r,\theta,z)}{C_T}=\frac{\partial X_{\mathrm{A}}}{\partial t}$$

Spherical coordinates

$$\frac{D_{\mathrm{A-B}}}{r^2}\left[\frac{\partial}{\partial r}\left(r^2\frac{\partial X_{\mathrm{A}}}{\partial r}\right)+\frac{1}{\sin\theta}\frac{\partial}{\partial\theta}\left(\sin\theta\frac{\partial X_{\mathrm{A}}}{\partial\theta}\right)+\frac{1}{\sin^2\theta}\frac{\partial^2 X_{\mathrm{A}}}{\partial\phi^2}\right]+\frac{\dot{r}_{\mathrm{A}}(r,\theta,\phi)}{C_T}=\frac{\partial X_{\mathrm{A}}}{\partial t}$$

Equations of Diffusion in the Presence of Fluid Flow (Constant Diffusivity, Constant Density)

Cartesian coordinates

$$D_{\mathrm{A-B}}\left(\frac{\partial^2 X_{\mathrm{A}}}{\partial x^2}+\frac{\partial^2 X_{\mathrm{A}}}{\partial y^2}+\frac{\partial^2 X_{\mathrm{A}}}{\partial z^2}\right)-\left(u_x\frac{\partial X_{\mathrm{A}}}{\partial x}+u_y\frac{\partial X_{\mathrm{A}}}{\partial y}+u_z\frac{\partial X_{\mathrm{A}}}{\partial z}\right)+\frac{\dot{r}_{\mathrm{A}}(x,y,z)}{C_T}=\frac{\partial X_{\mathrm{A}}}{\partial t}$$

Cylindrical coordinates

$$D_{\mathrm{A-B}}\left(\frac{\partial^2 X_{\mathrm{A}}}{\partial r^2}+\frac{1}{r}\frac{\partial X_{\mathrm{A}}}{\partial r}+\frac{1}{r^2}\frac{\partial^2 X_{\mathrm{A}}}{\partial \theta^2}+\frac{\partial^2 X_{\mathrm{A}}}{\partial z^2}\right)-\left(u_r\frac{\partial X_A}{\partial r}+\frac{u_\theta}{r}\frac{\partial X_A}{\partial\theta}+u_z\frac{\partial X_A}{\partial z}\right)+\frac{\dot{r}_{\mathrm{A}}(r,\theta,z)}{C_T}=\frac{\partial X_{\mathrm{A}}}{\partial t}$$

Spherical coordinates

$$\frac{D_{\mathrm{A-B}}}{r^2}\left[\frac{\partial}{\partial r}\left(r^2\frac{\partial X_{\mathrm{A}}}{\partial r}\right)+\frac{1}{\sin\theta}\frac{\partial}{\partial\theta}\left(\sin\theta\frac{\partial X_{\mathrm{A}}}{\partial\theta}\right)+\frac{1}{\sin^2\theta}\frac{\partial^2 X_{\mathrm{A}}}{\partial\phi^2}\right]-\left(u_r\frac{\partial X_{\mathrm{A}}}{\partial r}+\frac{u_\theta}{r}\frac{\partial X_{\mathrm{A}}}{\partial\theta}+\frac{u_\phi}{r\sin\theta}\frac{\partial X_{\mathrm{A}}}{\partial\theta}\right)+\frac{\dot{r}_{\mathrm{A}}(r,\theta,\phi)}{C_T}=\frac{\partial X_{\mathrm{A}}}{\partial t}$$

---

[a]In these equations, $\nabla^2=\frac{1}{r^2}\frac{\partial}{\partial r}\left(r^2\frac{\partial}{\partial r}\right)+\frac{1}{r^2\sin\theta}\frac{\partial}{\partial\theta}\left(\sin\theta\frac{\partial}{\partial\theta}\right)+\frac{1}{r^2\sin^2\theta}\left(\frac{\partial^2}{\partial\phi^2}\right)$

For turbulent fluid flow problems, in a formal sense, the corresponding equations are quite similar:

$$\nabla\cdot\rho\bar{\mathbf{U}}=0 \quad \text{(equation of continuity)} \tag{14}$$

and

$$\frac{\partial}{\partial t}(\rho\bar{\mathbf{U}})=-\mathbf{\nabla}\cdot\rho\bar{\mathbf{U}}\bar{\mathbf{U}}+\nabla\mu_{\mathrm{eff}}\mathbf{\nabla}\bar{\mathbf{U}}+\rho\mathbf{F}_b-\mathbf{\nabla}\bar{P} \quad \text{(equation of motion)} \tag{15}$$

where the overbar terms denote the time-smoothed quantities.† The basic difference is that μ_{eff}, the effective viscosity, is no longer a property value of the fluid, but is a strongly position-dependent quantity that is a function of the flow field within the system. As will be discussed in Section 6.1.1, in the solution of turbulent fluid flow problems, appropriate models of the turbulent viscosity in the form of auxilliary differential equations are needed.

The following additional comments should be made here:

The equations of continuity and motion have to be solved together, because in a physical sense both the conservation of matter and the conservation of momentum have to be satisfied.

In order to proceed with the solution of Eqs. (12) and (13) or (14) and (15), the body force field (F_b) may have to be specified. Thus, for electromagnetically driven flows, we need to solve subsidiary equations, such as Maxwell's equations; or for natural convection (i.e., buoyancy-driven) flows, we must know the temperature profiles.

In many instances, we may not be interested in the velocity field per se, but in the knowledge of the velocity field as an intermediate stage in the solution of a convective heat or mass transfer problem.

We rarely have to use the full set of the Navier–Stokes equations. Often, one may make simplifying assumptions that permit the elimination of the time derivative (in steady-state problems) or one or more of the spatial derivatives.

E. Heat Transfer

Conduction of heat in a solid may be represented by Fourier's law, which may be written as

$$\rho C_p \frac{\partial T}{\partial t} = \nabla k \nabla T + \dot{q} \tag{16}$$

where T is temperature,
C_p is the specific heat,
k is the thermal conductivity, and
$\dot{q}$ is the rate of heat generation (e.g., due to induction, chemical reactions, or a phase change).

The equivalent forms of Eq. (16) in Cartesian, spherical, and cylindrical coordinates are given in Table 2.2.

In general, when thermal energy is being transferred in fluids, in addition to

†Time smoothing, or averaging, is defined as $(1/t_0)\int_0^{t_0} F(t)\,dt$, where F is a variable, such as P or $\mathbf{U}$, that varies with time.

conduction, convection—caused by the bulk motion of the fluid—may also play an important, if not dominant, role. In a fluid undergoing bulk motion, the conservation of thermal energy may be expressed as

$$\underset{\text{(accumulation)}}{\rho C_p \frac{\partial T}{\partial t}} = \underset{\text{(convective transport)}}{-\rho C_p \mathbf{U} \cdot \nabla T} + \underset{\substack{\text{(conductive transport may}\\ \text{include turbulent transfer)}}}{\nabla k_{\text{eff}} \nabla T} + \underset{\text{(heat generation)}}{\dot{q}} \tag{17}$$

The equivalent forms of Eq. (17) in Cartesian, spherical, and cylindrical coordinates are also given in Table 2.2.

The principal difference between Eqs. (16) and (17) is the presence of the convective heat flow term ($\rho C_p \mathbf{U} \cdot \nabla T$) together with the fact that for a turbulent system the effective thermal conductivity (k_{eff}) is no longer a material property, but a function of the fluid flow field as well.

Clearly, in generating solutions for Eq. (17), the knowledge of the velocity field—that is, of $u(x, y, z)$—is a necessary prerequisite. In forced convection problems, the velocity field may be calculated independently, through the solution of Eqs. (12) and (13) or (14) and (15), and then one may proceed with the solution of Eq. (17), utilizing the velocity field now known.

The situation is somewhat more complicated in the case of natural convection problems where the driving force for the flow is provided by temperature gradients. Under these conditions the convective heat flow and the fluid flow equations have to be solved together.

F. Mass Transfer

1. DIFFUSION. For low concentration of the diffusing species and in the absence of convection, the conservation of the transferred species is given by Fick's law:

$$\frac{\partial C_{\text{A}}}{\partial t} = D_{\text{A-B}} \nabla^2 C_{\text{A}} + R_{\text{A}} \tag{18}$$

where $D_{\text{A-B}}$ is the binary diffusion coefficient,

C_{A} is the concentration of the diffusing species, and

R_{A} is the rate at which species A is being generated due to homogeneous chemical reaction (if we are concerned with a heterogeneous reaction, occurring at a surface, this would be introduced through the boundary conditions pertaining to that surface).

It should be noted here that Eq. (18) is applicable only when the diffusing species is present at low concentration levels and/or when there is equimolar counter-diffusion. When these conditions are not met, or for multicomponent systems, rather more complex relationships have to be invoked. For a full discussion of these considerations, the reader is referred to Bird, Stewart, and Lightfoot[5] or Szekely and Themelis.[6] Other complications may arise when diffusion takes place in porous media.[7]

It should be remarked that "pure diffusion" (i.e., in the absence of bulk motion) occurs relatively rarely in metallurgical practice, except in solids or in the pores of porous solid materials. Indeed, avoiding the establishment of convection currents in the experimental measurement of diffusion coefficients is a major practical problem.

2. CONVECTIVE MASS TRANSFER. In the presence of fluid motion, mass transfer takes place both by convection and by diffusion; and under these conditions the conservation of the diffusing species in a binary system may be written as

$$\underset{\text{(accumulation)}}{\frac{\partial C_A}{\partial t}} = \underset{\substack{\text{(convective}\\ \text{transport)}}}{\mathbf{U}\cdot\nabla C_A} + \underset{\substack{\text{(diffusive transport}\\ \text{including turbulent}\\ \text{component)}}}{\nabla D_{\text{eff}}\nabla C_A} + \underset{\substack{\text{(generation of A}\\ \text{due to chemical}\\ \text{reaction)}}}{R_A} \tag{19}$$

As in convective heat transfer problems, knowledge of the velocity field is required in order to solve Eq. (19); furthermore, for turbulent flow the parameter D_{eff} has to be deduced from the turbulent properties of the system. As before for forced convection, **U** may be determined independently from the diffusion process, while for natural convection the system of equations would become mutually coupled. [In case of diffusion in a thermally driven natural convection system, the heat flow and the fluid flow equations have to be solved together before Eq. (19) is tackled.]

In reacting systems a further complication arises through the fact that in general R_A, the rate of reaction, will be a function of the concentration of one or more reacting species. The rate expressions, which are often nonlinear, have to be obtained experimentally.

G. Electromagnetic Phenomena

Currently, widespread use is made in process metallurgy of equipment in which the flow of electric current is central to the process. Examples of such equipment are induction furnaces, submerged arc furnaces, and the electrolytic cells used to produce aluminum or magnesium. Some discussion of electromagnetic phenomena of relevance to modeling is therefore important here.

The currents are significant in that they may (1) generate electromagnetic forces that drive circulation of melts, (2) provide electrical heating, and (3) bring about electrochemical reactions.

The electromagnetic force appears as a body force in the Navier–Stokes equations (13) and (15). The electromagnetic force vector is due to the interaction of the current and magnetic fields and is given by

$$\mathbf{F}_b = \mathbf{J} \times \mathbf{B} \tag{20}$$

where **J** is the current density vector, and
B is the magnetic induction vector.

In dc systems Eq. (20) is used directly, since **J** and **B** do not vary with time. In ac systems, however, **J** and **B** are varying sinusoidally with a rapidity far too great for the fluid to keep up with. Consequently, whether the flow is laminar or turbulent, in the ac case, the product $\mathbf{J} \times \mathbf{B}$ is replaced by its time average value:

$$\mathbf{F}_b = \overline{\mathbf{J} \times \mathbf{B}} \tag{21}$$

Equation (13) or (15) must be coupled with appropriate boundary conditions on velocity (e.g., zero normal component of melt velocity at a solid wall). In addition the distribution of **J** and **B** throughout the melt must be determined in order to obtain the electromagnetic stirring force distribution. The way in which **J** and **B** are obtained depends on whether induced currents are significant (e.g., the induction furnace where all currents in the melt are induced) or insignificant (e.g., the Hall–Héroult cell). In the latter case the current density is obtained as follows. Since there are no current sources or sinks within the melt

$$\nabla \cdot \mathbf{J} = 0 \tag{22}$$

J is given by Ohm's law:

$$\mathbf{J} = \sigma \nabla \bar{\Phi} \tag{23}$$

where σ is the electrical conductivity, and
$\bar{\Phi}$ is the electric potential.

[Strictly, Eq. (23) should be modified to allow for the motion of the fluid. See Eq. (29)]. Subsequently

$$\nabla^2 \bar{\Phi} = 0 \tag{24}$$

which can be solved, with the potential specified at the bounding surfaces, to give the distribution of $\bar{\Phi}$, from which the distribution of **J** can be found by the application of Eq. (23). The Biot–Savart law can then be applied to calculate the magnetic induction **B** at all points in the melt due to current both within and without the melt. This is discussed in more detail in Chapter 8.

In the case where the currents arise partly or wholly from induction—and this is the case of interest in induction furnaces or induction stirrers—Maxwell's equations must be solved. For conditions of metallurgical interest these are

$$\nabla \times \mathbf{E} = -\frac{\partial \mathbf{B}}{\partial t} \tag{25}$$

$$\nabla \times \mathbf{B} = \mu_0 \mathbf{J} \tag{26}$$

$$\nabla \cdot \mathbf{B} = 0 \tag{27}$$

$$\nabla \cdot \mathbf{E} = 0 \tag{28}$$

where $\mathbf{E}$ is the electric field vector, and

μ_0 is the melt permeability (essentially that of vacuum $= 4\pi \times 10^{-7}$ in mks units).

The equations must be coupled with Ohm's law, written for a moving conductor:

$$\mathbf{J} = \sigma(\mathbf{E} + \mathbf{V} \times \mathbf{B}) \tag{29}$$

Under conditions of metallurgical interest the second term on the right of this equation is small and the normal form of Ohm's law [Eq. (23)] may be used.

Induced currents arise in ac operations in metallurgy, that is, in cases where currents, and subsequently **E** and **B** are varying sinusoidally. In these cases it is usual to employ an artifice developed by electrical engineers; this is to treat **J, E,** and **H** as phasors. In this approach the real (in the physical sense) **J**, **H**, and **E** are regarded as the real parts (in the mathematical sense) of complex quantities, that is,

$$\mathbf{J} \equiv \mathbf{J}_0 e^{j\omega t} \tag{30}$$

and so on

where $\mathbf{J}_0$ is a vector constant,

ω is the angular frequency (frequency $\times$ 2π), and

t is time.

Having embraced this concept, one is free to replace **J** with the phasor $\mathbf{J}_0 e^{j\omega t}$ in the various equations and similarly to replace **E** with $\mathbf{E}_0^{j\omega t}$ and **B** with $\mathbf{B}_0 e^{j\omega t}$. Subsequently the equations can be manipulated *in any way that does not entail multiplication of phasors.* The real and imaginary parts of the phasors remain separate, and at the end of the calculations one is free to take the real parts of the results as those having physical meaning. The virtue of the phasor approach is that the time derivative in Eq. (25) becomes $j\omega\mathbf{B}$ and the time dependence of the Maxwell equations is eliminated, at the expense of the complication stemming from the fact that **J, E,** and **B** are now complex. As mentioned above, the phasor approach is valid provided we refrain from taking products of phasors; such products result in the contribution of the imaginary parts (in both physical and mathematical senses) of the phasors to the real part of the product. For example, in using Eq. (21), the physically real force is obtained by time-averaging the products of the real parts of each phasor on the right, *not* by time-averaging the real part of the phasor product.

It should be noted that the specification of the real and imaginary parts of a phasor fixes the phase angle (ϕ) of the phasor and its magnitude, for example, through

$$\mathbf{J} \equiv \mathbf{J}_0(\cos \Phi + j \sin \Phi) \tag{31}$$

and enables the calculation of the real $[\mathbf{J}_0 \cos(\Phi + \omega t)]$ and imaginary $[\mathbf{J}_0 \sin(\Phi + \omega t)]$ parts of the phasor at any time (t) later. Consequently, phase calculations carried out for one instant enable calculation of time-averaged quantitites, such as $\mathbf{F}_b$ in Eq. (21).

Turning to the question of electrical heating, the local heat generation rate due to the current is given by

$$\dot{q} = \frac{J^2}{\sigma} \tag{32}$$

This would contribute to the heat generation term appearing above in Eq. (17). The question of electrochemical reactions is beyond the scope of this text, although it is touched upon in Chapter 8.

H. Other Relationships

Equations (12)–(19), together with the appropriate boundary conditions or the equivalent expressions given in Table 2.2, represent the key building blocks that may be used in the construction of mathematical models.

However, one frequently has to use some additional, subsidiary relationships in order to complete the statement of the problem. Some of these include the following:

In many problems one needs expressions describing the concentration, temperature, and possibly pressure dependence of the transport coefficients (i.e., the viscosity, thermal conductivity, and diffusion coefficients).

In reacting systems one needs an expression for the temperature and concentration dependence of the reaction rate (i.e., the rate expression).

In turbulent fluid flow systems one needs an appropriate turbulence model in order to compute the turbulent transport coefficients.

Some other relationships may be particularly important in connection with the boundary conditions:

Expressions for surface tension effects in determining a meniscus shape or surface tension–driven flows.

Empirical "friction factors"; heat transfer or mass transfer coefficients, particularly in the case of a "lumped parameter" approach, and the like.

The inexperienced reader may find the choice of the building blocks in the form of the transport equations quite bewildering. Indeed, considerable caution must be exercised in the construction of mathematical models for processes when using this approach. As with the case of model formulation through the control volume approach, the model preparation phase is crucial here in deciding the key simplifying assumptions (i.e., steady or unsteady state, the dimensionality of the problem, and the like) before proceeding with the selection of the actual equations.

2.3.3. Solution of the Equations

To solve the equations that have been obtained, one of two paths may be followed. An analytical solution may be sought to achieve an equation giving values of the dependent variable continuously with space and/or time. Alternatively, a numerical solution in which the differential equation is transformed into algebraic equations, and values of the dependent variable are determined at discrete intervals of space and/or time, may be undertaken. Of the two, an analytical solution is preferable but for most practical process problems is not achievable owing to the complexity of the mathematical equations and associated boundary or initial conditions. Under these circumstances the mathematics have to be simplified considerably to make an analytical solution possible, but this, of course, usually results in model predictions that are well removed from reality. This is not to suggest that the analytical solution has no place in a mathematical model. For simple problems, an analytical solution may be achievable, and even for more complicated situations it may be employed to check a numerical solution under special or limiting conditions. For example, a numerical solution for the temperature distribution in a slab being heated in a reheating furnace could be checked with an analytical solution based on constant thermophysical properties, symmetrical heating, and a constant surface heat transfer coefficient. Such a solution can be found in any undergraduate heat transfer textbook.[9]

The significant difference between an analytical solution and a numerical one is that the former presents a functional relationship (usually an algebraic equation) between the dependent and independent variables. As its name implies, the numerical solution is precisely the numerical value of the dependent variable at particular values of the independent variables. The analytical solution is thus available continuously, whereas the numerical solution is available only at a finite number of "nodes," usually arranged in a "mesh," as will be described in Section 2.3.3.D.

A. The Concept of Scaling

Before proceeding with the generation of analytical or numerical solutions, a very useful intermediate step is available through the concept of scaling. When the governing equations are made dimensionless *in the proper way*, the dimensionless groups that emerge may serve as an excellent semiquantitative indication of the overall system behavior that can be expected. Furthermore, by the proper choice of these dimensionless parameters, useful order-of-magnitude (or even better) estimates regarding relationships among the key process parameters may be obtained. This may be illustrated by a few simple examples.

EXAMPLE 1

Consider a steel slab of thickness L that is initially at a temperature T_i and for subsequent time is losing heat by radiation to the surroundings having an absolute temperature T_E. This problem may be formulated as

$$\rho C_p \frac{\partial T}{\partial t} = k \frac{\partial^2 T}{\partial y^2}, \qquad 0 \leqslant y \leqslant L \tag{33}$$

with

$$T = T_i \qquad \text{at} \quad t = 0 \tag{34}$$

$$-k \frac{\partial T}{\partial y} = \varepsilon\sigma(T^4 - T_E^4) \qquad \text{at} \quad y = 0, \qquad y = L \tag{35}$$

Solution. This equation will have to be solved numerically because of the nonlinearity in the boundary conditions. However, upon introducing the following dimensionless parameters:

$$\beta = \frac{y}{L} \quad \text{(dimensionless length)} \tag{36}$$

$$\tau = \frac{t\alpha}{L^2} \quad \text{(dimensionless time)} \tag{37}$$

$$\alpha = \frac{k}{\rho C_p} \tag{38}$$

$$\tilde{T} = \frac{T}{T_i} \quad \text{(dimensionless temperature)} \tag{39}$$

Eq. (33) becomes

$$\frac{\partial \tilde{T}}{\partial \tau} = \frac{\partial^2 \tilde{T}}{\partial \beta^2}, \qquad 0 \leqslant \beta \leqslant 1 \tag{40}$$

where τ is the Fourier number, and
$t\alpha/L^2$ is the time scale.

A rough estimate of the time required to attain steady state (i.e., for the slab to cool down to T_E from T_i) may now be obtained by setting

$$\frac{t\alpha}{L^2} \sim 1, \qquad \text{that is,} \quad t_{eq} \approx \frac{L^2}{\alpha}$$

EXAMPLE 2

Similar considerations can be applied to molecular diffusion and to viscous flow. For a transient diffusion problem, in the absence of convection the time scale would be

$$\frac{Dt}{L^2}$$

where L is the characteristic length, and
D is the diffusion coefficient.

Solution. Thus, t_{eq}, the time for the attainment of equilibrium, is

$$t_{eq} \sim \frac{L^2}{D}$$

EXAMPLE 3

Similarly, for the decay of a laminar velocity field in a small agitated pool after stirring has been discontinued, the time scale is vt/L^2

where v is the kinematic viscosity.

Solution. It follows that the time of inertial decay may be estimated as

$$\frac{vt}{L^2} \sim 1, \qquad \text{that is,} \quad t_d \sim \frac{L^2}{v}$$

Similar considerations may be applied to more complex convection problems. Thus, the general statement of the equation of motion in the presence of buoyancy and electromagnetic forces may be written as

$$\rho \frac{\partial \mathbf{U}}{\partial t} = \mathbf{U} \cdot \nabla \mathbf{U} + \mu \nabla^2 \mathbf{U} - \nabla P + \mathbf{J} \times \mathbf{B} + \rho \mathbf{g} \tag{41}$$

By making these equations dimensionless, the order of magnitude, and hence the importance of,

$$\mathbf{U} \cdot \nabla \mathbf{U} \quad \text{(inertial term)} \qquad \text{and} \qquad \mu \nabla^2 \mathbf{U} \quad \text{(viscous term)}$$

or

$$\mathbf{J} \times \mathbf{B} \quad \text{(electromagnetic force field)} \qquad \text{and} \qquad \rho\mathbf{g} \quad \text{(buoyancy field)}$$

can be assessed. It may be shown, furthermore, that when electromagnetic forces dominate,

$$\mathbf{U} \sim \mathbf{J}^2 \quad \text{for laminar flow}$$

and

$$\mathbf{U} \sim \mathbf{J} \quad \text{for turbulent flow}$$

In addition, by assigning proper values to the characteristic parameters one may obtain order-of-magnitude estimates of the key process variables. A good discussion of these issues can be found in several references.[10–12]

The principles of numerical methods are set out briefly below. For a more detailed treatment, the interested reader is referred to many excellent textbooks on the subject.[13–17]

B. Finite-Difference Approximations

The mathematical basis for numerical methods can be seen by considering the definition of a first derivative:

$$\left.\frac{dy}{dx}\right|_x = \lim_{\Delta x \to 0} \left[\frac{y|_{x+\Delta x} - y|_x}{\Delta x}\right] \tag{42}$$

Since the dependent variable is only available at nodes separated by finite intervals of space or time, we have to approximate a derivative such as the above by, for example,

$$\left.\frac{dy}{dx}\right|_{x_i} \simeq \frac{y_{i+1} - y_i}{\Delta x} \tag{43}$$

where x_i and y_i are the independent and dependent variables, respectively, at the node designated i, and $\Delta x = x_{i+1} - x_i$. This finite-difference approximation is frequently called the "forward difference." Other common finite-difference approximations to derivatives are

$$\left.\frac{dy}{dx}\right|_{x_i} \simeq \frac{y_{i+1} - y_{i-1}}{2\Delta x} \quad \text{("central difference")} \tag{44}$$

$$\left.\frac{dy}{dx}\right|_{x_i} \simeq \frac{y_i - y_{i-1}}{\Delta x} \quad \text{("backward difference")} \tag{45}$$

Similarly, the approximation to a second derivative can be written as

$$\left.\frac{d^2 y}{dx^2}\right|_{x_i} \simeq \frac{y_{i+1} - 2y_i + y_{i-1}}{(\Delta x)^2} \tag{46}$$

All these are derivable by simple manipulation of Taylor series expansions of y_i and x_i and are tabulated in texts on numerical methods (e.g., Abramowitz and Stegun).[13] Note that these are approximations; they contain a "truncation error" arising from taking a finite number of terms in the Taylor series expansion. The truncation error diminishes rapidly as Δx is reduced, however.

C. Ordinary Differential Equations

In some process models, the dependent variable is sought only as a function of a single independent variable. For example, a model of a rotary kiln may have been formulated to predict concentration and temperature profiles in the axial direction only at steady state. Another example is a model of thin steel strip being cooled by water jets on a run-out table in which the axial temperature profile of the strip is to be predicted at steady state. In these cases, one or more first-order ordinary differential equations will be formulated from the balances described earlier.

Consider the first-order ordinary differential equation

$$\frac{dy}{dx} = f(x, y) \tag{47}$$

with a boundary condition $y = y_0$ at $x = x_0$. To solve this equation, from the discussion above, the first requirement is to divide the range of the independent variable into discrete intervals of size Δx. For now, Δx will not vary with x. The solution, then, will be to obtain the values of $y(y_1, y_2, y_3, \ldots, y_n)$ at the specified discretely separated values of $x(x_1, x_2, \ldots, x_n)$. This is done by starting at x_0 (where $y = y_0$) and "marching" forward with a "step size" of Δx.

This can be accomplished most simply by approximating the derivative with the forward finite-difference approximation as follows:

$$\frac{y_1 - y_0}{\Delta x} = \left.\frac{dy}{dx}\right|_{x_0} = f(x_0, y_0) \tag{48}$$

Since x_0 and y_0 are known, $f(x_0, y_0)$ is known, and we can immediately calculate y_1 by rearranging the above: $y_1 - y_0 = \Delta x f(x_0, y_0)$. Subsequently, we can find y_2:

$$y_2 = y_1 + \Delta x f(x_1, y_1) \quad \text{and so on}$$

This is Euler's method, which has the advantage of simplicity but is inaccurate unless Δx is kept so small that the pace with which the solution marches forward is very slow.

Euler's method is one example of a class of methods called Newton-Coates methods, which have the general form

$$y_{i+1} = y_i + \Delta x \phi(x_i, y_i) \tag{49}$$

where $\phi(x_i, y_i)$ is called an "increment function." Perhaps the most widely used Newton-Coates method is the Runge–Kutta method where

$$\phi(x_i, y_i) = \frac{1}{6}[k_1 + 2k_2 + 2k_3 + k_4] \tag{50}$$

and

$$k_1 = f(x_i, y_i), \qquad k_2 = f\left(x_i + \frac{\Delta x}{2}, y_i + \frac{\Delta x k_1}{2}\right)$$

$$k_3 = f\left(x_i + \frac{\Delta x}{2}, y_i + \frac{\Delta x k_2}{2}\right), \qquad k_4 = f(x_i + \Delta x, y_i + \Delta x k_3)$$

Note that this method proceeds by first taking a "half step" within each step.

An alternative class of methods is known as predictor–corrector methods, where

$$y_{i+1} = y_i + \Delta x \phi(x_{i+1}, y_{i+1}, x_i, x_{i-1}, y_{i-1}, \ldots) \tag{51}$$

These methods are interesting in that, in order to calculate the increment function to obtain y_{i+1}, we need to know what the value of y_{i+1} is! The solution to this dilemna is that the equation is applied "iteratively." We start with a crude approximation to y_{i+1}, which is then inserted in the right-hand side of Eq. (51), and (it is hoped) an improved value is calculated. Further refinement can be achieved by additional applications of this "corrector" equation. In order to obtain our crude approximation, we first use a "predictor" equation of the form

$$y_{i+1} = y_i + \Delta x f(x_i, y_i, x_{i-1}, y_{i-1}, \ldots) \tag{52}$$

Each step in marching forward consists of the application of the predictor equation followed by one or more applications of the corrector. There are several predictor-corrector methods. Usually they are more difficult to program than the Runge-Kutta method but require less computer time to run.

Equations higher than first order can be handled by simple substitutions, for example,

$$\frac{d^2 y}{dx^2} + \frac{dy}{dx} = f(x, y) \tag{53}$$

can be reduced to two simultaneous ordinary differential equations:

$$\frac{dy}{dx} = z \tag{54}$$

$$\frac{dz}{dx} = f(x, y) - z \tag{55}$$

which can be marched forward simultaneously starting with an initial condition

$$\frac{dy}{dx} = z = z_0 \qquad \text{and} \qquad y = y_0 \quad \text{at} \quad x = x_0 \tag{56}$$

Note that a complication can arise for higher order equations in that the boundary conditions may be specified at more than one point. An example of such a "boundary-value problem" would be the boundary conditions

$$\frac{dy}{dx} = a \qquad \text{at} \quad x = x_a \tag{57}$$

$$y = b \qquad \text{at} \quad x = x_b \tag{58}$$

for the above second-order equation. The solution of boundary value problems is discussed in Section 2.3.3.I.

D. Partial Differential Equations

The equations resulting from the balance described in Section 2.3.2.B frequently contain two or more independent variables, such as time and spatial position. A common example is the cooling of a slab during rolling, in which the unsteady-state conduction of heat within the slab is described by the second-order partial differential equation

$$k\frac{\partial^2 T}{\partial z^2} = \rho C_p \frac{\partial T}{\partial t} \tag{59}$$

A model based on this equation will be discussed in some detail in Section 2.4.1. Equation (59) is a parabolic-type partial differential equation that can be written more generally as

$$\frac{\partial^2 u}{\partial x^2} = \frac{\partial u}{\partial t} \tag{60}$$

Other types of partial differential equation are the elliptic:

$$\frac{\partial^2 u}{\partial x^2} + \frac{\partial^2 u}{\partial y^2} = 0 \tag{61}$$

and the hyperbolic:

$$\frac{\partial^2 u}{\partial x^2} + \frac{\partial^2 u}{\partial y^2} = \frac{\partial^2 u}{\partial t^2} \tag{62}$$

The elliptic partial differential equation arises in steady-state problems such as heat flow through the copper mold of a continuous-casting machine operating at steady state.

The boundary conditions for these equations are often nonlinear, as in the case of radiative heat flow to or from a slab surface; or the properties, such as thermal conductivity or diffusion coefficient, are functions of the dependent variable. In other cases a change of phase and a phase-change temperature interval of finite size (e.g., difference between liquidus and solidus temperatures) may be involved. These complications usually make it necessary to seek numerical solutions that most frequently are achieved by applying the finite-difference method. As in the case of the ordinary differential equations, this method is based on the discretization of space and time into steps of Δx, Δy, and Δt, and the application of finite-difference approximations to derivatives.

The mathematical underpinning of the finite-difference method can be grasped easily by considering a two-dimensional $x-y$ space that has been divided into a grid of nodes, as shown in Figure 2.2. The nodes are identified by the indices i and j. Employing the finite-difference approximations already described, the derivatives in Eqs. (60)–(62) can be approximated by differences

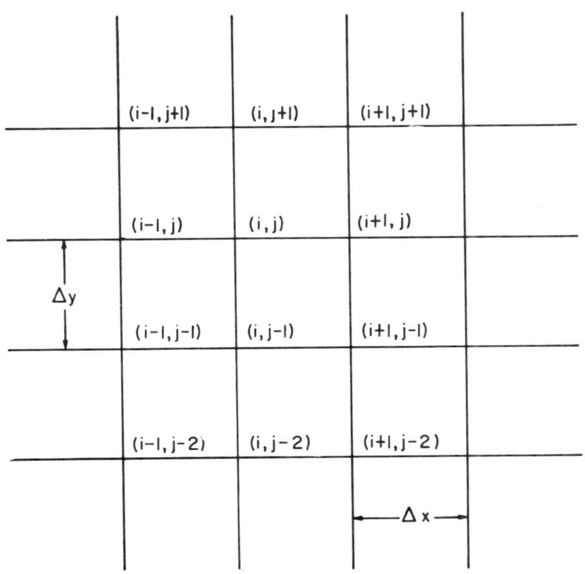

FIG. 2.2. Subdivision of a two-dimensional space into a finite difference grid.

between values of the dependent variable u at the nodes, for example,

$$\frac{\partial^2 u}{\partial x^2} \approx \frac{u_{i+1,j} - 2u_{i,j} + u_{i-1,j}}{(\Delta x)^2} \tag{63}$$

Such finite-difference approximations can be substituted appropriately into the partial differential equation to be solved—say, Eq. (61)—and the values of $u_{i,j}$ calculated throughout the nodal grid (Figure 2.2).

The mechanics of the application of the finite-difference method are best seen by examining a specific problem involving the frequently encountered parabolic-type equation. Consider a horizontal cooling slab of thickness Z in which the heat fluxes $\dot{q}_0$, from the top and bottom surfaces are equal. Far from the edges of the slab, heat flow is dominated by conduction in the through-thickness (z) direction only; and the temperature distribution is governed by Eq. (59) together with the following initial and boundary conditions:

$$t = 0, \qquad 0 < z < Z/2, \qquad T = T_0 \tag{64}$$

$$z = 0, \qquad t > 0, \qquad -k\frac{\partial T}{\partial z} = 0 \quad \text{(centerline)} \tag{65}$$

$$z = \frac{Z}{2}, \qquad t > 0, \qquad -k\frac{\partial T}{\partial z} = \dot{q}_0 \quad \text{(surface)} \tag{66}$$

To solve Eq. (59) several formulations of the finite-difference method will be considered. However, for each, the slab will be subdivided into intervals of Δz and Δt, as shown in Figure 2.3. Note that nodes are located at the centerline and surface of the slab; the indices, i and n, are employed to identify space and time intervals, respectively.

E. Explicit Finite-Difference Method

The finite-difference methods may be classified according to the manner in which the second-order spatial derivative, $\partial^2 T/\partial z^2$, is approximated. In the explicit formulation, $\partial^2 T/\partial z^2$ is expressed in terms of known temperatures at the present time step identified by the index n in Figure 2.3. Referring to the discretized slab in Figure 2.3, this means expressing $\partial^2 T/\partial z^2$ in terms of $T_{i-1,n}$, $T_{i,n}$, and $T_{i+1,n}$.

From Eq. (46),

$$\frac{\partial^2 T}{\partial z^2} \simeq \frac{T_{i+1,n} - 2T_{i,n} + T_{i-1,n}}{(\Delta z)^2} \tag{67}$$

From Eq. (43),

$$\frac{\partial T}{\partial t} \simeq \frac{T_{i,n+1} - T_{i,n}}{\Delta t} \tag{68}$$

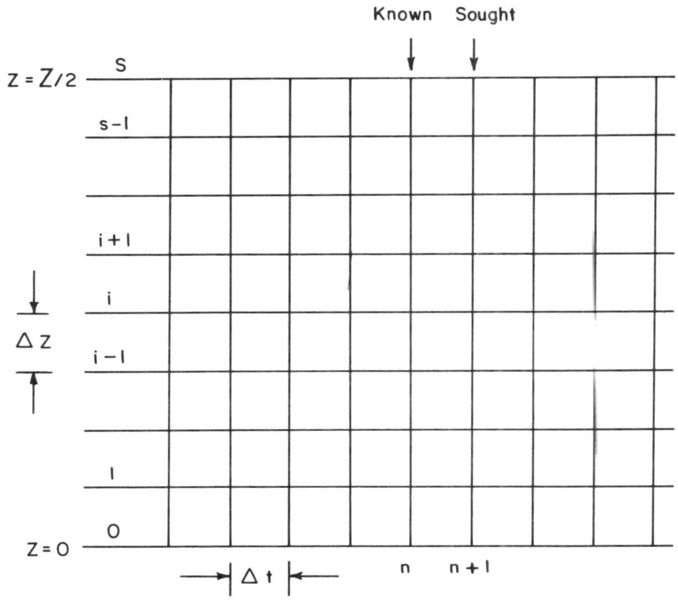

FIG. 2.3. Subdivision of a slab thickness and time into discrete intervals for finite difference calculations.

Substituting into Eq. (59) and rearranging,

$$T_{i,n+1} = T_{i,n} + \frac{k\Delta t}{\rho C_p(\Delta z)^2}[T_{i-1,n} - 2T_{i,n} + T_{i+1,n}] \tag{69}$$

Equation (69) can be solved *explicitly* to obtain $T_{i,n+1}$ for each *interior* node because the values of $T_{i-1,n}$, $T_{i,n}$, and $T_{i+1,n}$ are known. However, Eq. (69) does not apply to the boundaries, and separate equations have to be derived. At the surface boundary, this can be accomplished by defining a fictitious node, $s + 1$, and thereby making the surface artificially an interior node. Then the temperature of the fictitious node can be determined by substituting a central-difference approximation for the derivative in Eq. (66). After rearrangement,

$$T_{s+1} = T_{s-1} - \frac{2\Delta x\dot{q}_0}{k} \tag{70}$$

Substituting Eq. (70) into Eq. (69) where $i = s$ yields the following:

$$T_{s,n+1} = T_{s,n} + \frac{2k}{\rho C_p}\frac{\Delta t}{(\Delta z)^2}\left[T_{s-1,n} - T_{s,n} - \frac{\Delta z}{k}\dot{q}_C\right] \tag{71}$$

Similarly, below the centerline of the slab the temperature of the node -1 is determined by substitution of a central difference into Eq. (65) to yield

$$T_{-1} = T_1 \tag{72}$$

Substitution of Eq. (72) into Eq. (69) with $i = 0$ gives

$$T_{0,n+1} = T_{0,n} + \frac{2k}{\rho C_p} \frac{\Delta t}{(\Delta z)^2} [T_{1,n} - T_{0,n}] \tag{73}$$

Thus, at each successive time step, Eqs. (69), (71), and (73) are solved explicitly to obtain temperatures at the next time step. In this way, the computation marches through time, sweeping out the temperature distribution behind it.

It is important to note that there is a restriction on the choice of the size of the distance and time steps. It can be shown that for a stable solution to be achieved the following condition must hold:

$$\frac{k}{\rho C_p} \frac{\Delta t}{(\Delta z)^2} < \frac{1}{2} \tag{74}$$

Eq. (74) applies to one-dimensional heat conduction and an interior node. To obtain the stability requirements for other situations, the interested reader is referred to Kreith and Black.[9] From Eq. (74), the explicit finite-difference method is seen to have the disadvantage that the use of small space intervals requires small time steps and more computational effort.

F. Implicit Finite-Difference Methods

In the implicit method, the second-order spatial derivative is expressed in terms of unknown temperatures at the next time level, $n + 1$. Thus, for the ith interior node, the finite-difference equation is the same as Eq. (69), but the index n is replaced by $n + 1$.

In the simple implicit and explicit methods, the time derivative in Eq. (59) is approximated by a simple forward difference. The error associated with this approach can be reduced by performing a central difference at the midpoint of the time interval, $(n + 1/2)\Delta t$, which is the basis of the Crank–Nicolson method. The spatial derivative at the midpoint then is approximated by

$$\frac{\partial^2 T}{\partial z^2} = \frac{T_{i+1,n} - 2T_{i,n} + T_{i-1,n}}{2(\Delta z)^2} + \frac{T_{i+1,n+1} - 2T_{i,n+1} + T_{i-1,n+1}}{2(\Delta z)^2} \tag{75}$$

When these implicit approximations are substituted in the partial differential equation, $T_{i,n+1}$ cannot be obtained explicitly from $T_{i+1,n}$, $T_{i,n}$, and $T_{i-1,n}$ as before but must be obtained by solving a set of linear simultaneous equations where $T_{i,n+1}$ depends on $T_{i+1,n+1}$ and $T_{i-1,n+1}$. These linear algebraic equations take on a tridiagonal form:

$$
\begin{aligned}
b_1 T_1 + c_1 T_2 \qquad\qquad\qquad &= d_1 \\
a_2 T_1 + b_2 T_2 + c_2 T_3 \qquad &= d_2 \\
a_3 T_3 + b_3 T_3 + c_3 T_3 &= d_3 \\
a_m T_{m-1} + b_m T_m &= d_m
\end{aligned}
$$

which can be solved by Gauss elimination or inversion of the matrix of coefficients. However, it is more usual to exploit the "sparseness" (the fact that most of its elements are zero) of this matrix and to make use of a fast algorithm from Thomas[14] to solve the equations.

When applied to a problem with one spatial dimension, the implicit methods described above yield a tridiagonal set of equations. If the problem has two or three spatial dimensions, then a pentadiagonal or septadiagonal set arises, and these are much more difficult to handle. In these cases the alternating-direction technique is used, in which, for example, in a problem in two spatial dimensions the equations are solved explicitly in the x direction and implicitly in the y direction. On the next step the equations are solved explicitly in the y direction and implicitly in the x direction. The technique may be extended to three space dimensions.

The great virtue of the implicit techniques is that the value of Δt is not limited by the stability criterion imposed on the explicit method. Therefore, the implicit techniques typically use much less computer time, although they are more difficult to program.

G. Nonlinear Equations

Numerical techniques have their major utility when the partial differential equation is nonlinear. An example would be unsteady-state conduction where the thermal conductivity, k, is a function of temperature. The simple approach here is to use an explicit method. The nonlinearity is then not a problem, since the required nonlinear terms are all explicitly known. A more sophisticated approach is to use an implicit method but to remove the nonlinearity by linearization about the old temperature, for example,

$$
k(T)\frac{\partial^2 T}{\partial z^2} \approx k(T_{i,n})\left[\frac{T_{i+1,n+1} - 2T_{i,n+1} + T_{i-1\ n+1}}{2(\Delta z)^2} + \frac{T_{i+1,n} - 2T_{i,n} + T_{i-1,n}}{2(\Delta z)^2}\right] \tag{76}
$$

This approach is satisfactory if the dependence of k on T is weak or Δt is kept small. Larger values of Δt can be used if the step is iterated. In other words, the above equation is employed one or more additional times with $k(T_{i,n})$ replaced by $0.5[k(T_{i,n}) + k(T_{i,n+1})]$ after the first iteration, the value for $T_{i,n+1}$ being that obtained from the previous iteration.

H. Elliptic Equations and Relaxation Methods

An example of an elliptic problem is the calculation of a steady-state temperature distribution in a body with heat generation, such as a busbar carrying electric current to a Hall–Heroult cell.

Consider a two-dimensional problem of this nature:

$$\frac{\partial^2 T}{\partial x^2} + \frac{\partial^2 T}{\partial y^2} = -\frac{\dot{q}}{k} \tag{77}$$

where $\dot{q}$ is the specified heat generation rate. Boundary conditions might be $T = T_0$ at the boundaries of the object. If the solution is obtained numerically, it will consist of a set of T values at the nodes of a mesh (assumed here to be rectangular). Let i and j be the indices of a node. The usual finite-difference approximations then lead to

$$\frac{T_{i+1,j} - 2T_{i,j} + T_{i-1,j}}{(\Delta x)^2} + \frac{T_{i,j+1} - 2T_{i,j} + T_{i,j-1}}{(\Delta y)^2} = \frac{\dot{q}}{k} \tag{78}$$

There is one such set of equations for each internal point, and the equation $T_{i,j} = T_0$ for each boundary point. Hence, we have a set of linear simultaneous equations that could be solved by inversion of the matrix of coefficients. It is more usual, however, to use an iterative approach, for example, the set of equations

$$\frac{T_{i+1,j} - 2T_{i,j} + T_{i-1,j}}{(\Delta x)^2} + \frac{T^*_{i,j+1} - 2T_{i,j} + T^*_{i,j-1}}{(\Delta y)^2} = \frac{\dot{q}}{k} \tag{79}$$

can be solved, where the $*$ indicates an "old value" obtained on the previous iteration. Such values are merely numbers, and then only $T_{i+1,j}$, $T_{i,j}$ and $T_{i-1,j}$ are unknowns. Such a set of equations can be readily manipulated into tridiagonal form and solved by Thomas' algorithm.

Frequently, such a "relaxation" scheme fails to "converge," that is, the values obtained after each iteration are further from a true solution than are the old values. Techniques (e.g., underrelaxation[14]) are available for circumventing this problem and speeding convergence. The achievement of convergence is usually determined empirically, that is, the equations are iterated until the solution fails to change significantly on further iteration.

I. Boundary Value Problems

The boundary value problem described in Section 2.3.3.C is analogous to the solution of an elliptic partial differential equation. The usual finite differences are employed to obtain a linear algebraic equation for each internal point in terms of its neighbors. Equations are written for the dependent variable at the boundaries, and there results a set of simultaneous linear equations that are amenable to Thomas' algorithm.

J. The Finite-Element Method

The finite-element method is another numerical technique that has traditionally been used in stress analysis but that is now also finding application in other areas, such as heat transfer. The mathematical basis for the finite-element method has its origin in the classical variational technique. In this method the differential equation and boundary conditions are converted to a functional and the problem reduces to seeking functions that extremize the functional. The Ritz method that is the forerunner of the finite-element method provided a means of obtaining an approximate solution to the variational problem. Trial functions with unknown adjustable parameters are substituted into the functional, which is in turn differentiated with respect to the adjustable parameters. The latter are obtained by solving the resulting equations. The finite-element method is a special case of the Ritz method and employs specific interpolating polynomials in place of trial functions. Other methods have also been developed to permit derivation of finite-element equations directly from the differential equations; they are the method of weighted residuals, the direct approach, and the energy balance approach. These methods are more versatile than the Ritz-type approach because they do not require the formulation of a functional, which is often difficult.

Irrespective of the method used to obtain the assemblage of finite-element equations, the basic steps involved in applying the method are the same. The first step is to divide the continuum into elements. Triangular and rectangular elements are widely employed for two-dimensional domains, while tetrahedra or general hexahedra are used to span three-dimensional spaces. The next step is to choose the type of interpolating polynomial to represent the variation of the field variable over an element. The degree of polynomial chosen is determined by the number of nodes assigned to an element, the tensor order of the field variable, and the continuity requirements imposed along boundaries. These criteria can be discerned from the functional or integral equations generated by the different methods. To guarantee convergence, the variable and its partial derivative appearing in the integral equations must be continuous within the element, a requirement that dictates the minimum order of an admissible polynomial. Second, all uniform states of the field variable and its partial derivative up to the highest order in the equations should be represented by the polynomial. These criteria determine whether the element is compatible. The matrix equations for an element can then be determined using the polynomials by one of the four methods described earlier: (1) Ritz-type, (2) weighted residuals, (3) direct approach, or (4) energy-balance approach. The unknowns in the resulting equations are the variable and/or its derivatives at the nodes. The individual element equations are "assembled" to form a single matrix equation expressing the behavior of the "assemblage," modified to account for boundary conditions and solved using a standard matrix solver for the unknown values of the field variable at the nodes.

It is important to note that the finite-element method, unlike the finite-difference or weighted residual methods, was initially developed for problems involving only spatial variation of the variable. For time-dependent field problems the finite-element method has been traditionally combined with appropriate finite-difference schemes to accomplish time stepping, although a few investigators have attempted to use finite-elements in time. The different finite-difference time-stepping schemes are distinguished by the number of previous time-step evaluations required in the algorithm. The commonly used algorithms are the two-level and three-level schemes. Despite the necessity of adulterating the finite-element method with finite-difference schemes for solving transient problems, the method has distinct advantages over the "pure" finite-difference methods. The finite-element method can be applied with greater ease and accuracy for problems involving complex geometry. Curved-sided isoparametric elements that utilize geometric transformations have been developed expressly for this purpose. From the point of view of cost, the finite-difference methods are more attractive, both in CPU time and virtual memory costs. It should be borne in mind, however, that the greatest cost disparity arises in the core storage requirements and associated virtual memory costs. The most commonly cited drawback of the finite-element method related to this aspect of the technique is the extensive effort required to generate the mesh and interelement connectivity. This drawback is rapidly being overcome due to the availability of software capable of automatic mesh generation for any general shape. The continued widespread use of the finite-element method will be determined in part by the success of developments directed at cost optimization; it will also depend on the ability of the methods to accurately model nonlinear material behavior, crack tip propagation, and other complex phenomena. Many texts are available for the interested reader.[18–20]

The foregoing treatment of numerical methods has concerned itself only with the solution of ordinary and partial differential equations. It should be recognized that the mathematical modeler may have occasion to use other numerical techniques, for example, numerical integration, interpolation, location of zeros of functions, and optimization. It should be recognized that most computer systems will already have available software packages to handle routine numerical techniques. While the blind use of such software can be dangerous, their use after reading the relevant manuals or referring to specialized texts (such as those mentioned earlier) can usually save a great deal of time in writing and correcting programs. Furthermore, most modern large computer systems have access to commercial programs developed for the solution of large systems of nonlinear coupled partial differential equations. Such programs are designed to provide the user with great flexibility with regard to boundary conditions, variation of properties, irregular geometry, and so on. Examples of such programs are ANSYS, PISCES, ABACUS, PR3D, ELLPACK, and PHOENIX. For very complex problems the use of such programs can be advantageous.

2.3.4. Validation of the Model

The fourth stage in the development of a mathematical model, its validation, is frequently the most difficult. At the same time, the importance of model validation cannot be overemphasized. It should be apparent by now that model building is based on approximations and assumptions that necessarily cause the model to be removed somewhat from reality. Therefore, the model must be checked against the real world to determine its adequacy in making predictions that are accurate enough to be useful. In other words, measurements must be made of temperature, composition, velocity, pressure, and so on in the process for comparison to model predictions. The mathematical model cannot be applied to meeting project objectives until this stage has been completed successfully. In many cases poor agreement between measurements and predictions requires that the model be changed and rechecked; and this iteration may have to be repeated several times.

This is not to say that if model predictions do not agree with measurements, the model is invalid. Particularly under industrial conditions, measurements also can be in error. For example, in one actual industrial experiment, the purpose of the trial was to check the ability of a mathematical model to predict the pool depth in a continuous-casting machine producing steel billets (the model is described in Chapter 7). A tungsten pellet with a radioactive tracer (^{60}Co) sealed inside was dropped into the liquid pool during casting, and at the same time a tungsten marker was frozen into the solidifying shell at the meniscus. The free-falling tungsten pellet, having a density greater than that of steel, was expected to fall through the pool and settle on the bottom. After the test billet had cooled, the distance between the two tungsten pellets was measured and found to be about 4 m, which was significantly shorter than the value predicted by the model. Some light was shed on the discrepancy when it was learned that the tungsten pellet was made originally from metal powder and contained a residual porosity of several percentage points. Apparently, the gas in these pores escaped from the pellet as it heated in the liquid steel; but the gas was trapped by the steel freeze encapsulating the pellet (the pellet was at room temperature when dropped into the pool). In this way the effective density of the pellet was reduced so that theoretically it could achieve neutral buoyancy. Depending on the quantity of gas evolved and its volume, the pellet would not necessarily sink rapidly to the bottom of the pool. To check this hypothesis a tantalum pellet made from wrought stock was dropped into the pool at the same time as another tungsten pellet. The tantalum fell 2 m farther, somewhat closer to the model predictions.

2.3.5. Application of the Model

The final stage in model development is applying the model to meet the project objectives. Even though the model has been validated, caution should still be exercised in interpreting and using the model predictions.

2.4. EXAMPLES OF DEVELOPMENT OF MATHEMATICAL MODELS

2.4.1. Temperature Distribution in a Steel Slab during Hot Rolling

Consider the practical problem of determining the temperature distribution in a steel slab during hot rolling to plate or strip in a mill, schematically shown in Figure 2.4. This problem is of considerable importance because the temperature must be controlled to ensure that the steel, as it passes through the various rolling stands, consists either of wholly the single austenite phase or of dual phases in a controlled rolling schedule. To aid in the characterization of the steel temperature, a mathematical model can be an invaluable tool.

Thus, the objective of the work is the quantitative determination of the temperature distribution in a steel slab as a function of several variables, including initial slab thickness, reheat temperature, design of the rolling mill (distance between rolling stands, size of rolls, and characteristics of scale-breakers, which are high-pressure water jets), and the rolling schedule. The development of a mathematical model to achieve this objective must be viewed favorably because a model provides the ability to predict slab temperatures off line. As shall be seen, however, the development of the model will require the measurement in plant of heat flow from the slab surface to the rolls, scale breakers, and surroundings.

In formulating the model, the first consideration is the possibility of neglecting any of the directions of heat flow. In particular, the question of heat flow across the slab width, from midface to edge, needs to be addressed. If this transverse heat flow is unimportant owing to uniform heat transfer over the width of the slab, and if the corners of the slab are not of concern, this component of heat flow can be neglected. The problem is then reduced to one of two-dimensional heat flow in a longitudinal plane passing through the slab, as shown in Figure 2.5. It is unlikely that heat flow can be neglected in the through-thickness (z) direction because the temperature gradients will be large, at least until the slab has been reduced considerably in thickness. This aspect can be checked, of course, by the Biot modulus, which is a measure of the relative magnitude of the internal resistance to heat flow (L/k) compared to the external resistance ($1/h$).

A volume element within the slab is next defined as shown in Figure 2.5. Note that the element consists of an infinitesimal area having dimensions of Δz by Δy

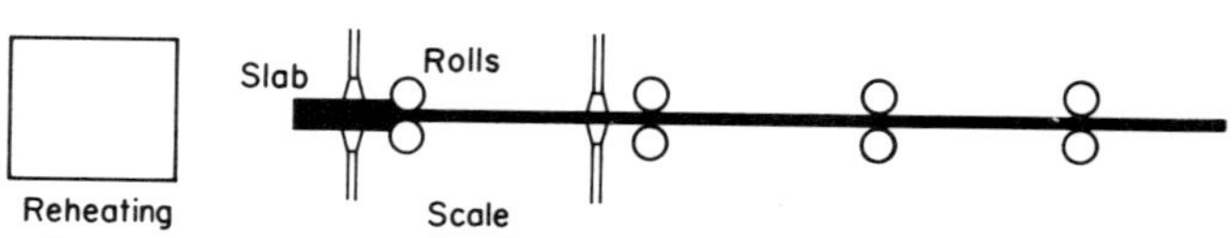

FIG. 2.4. Schematic drawing of a slab passing through a hot rolling mill.

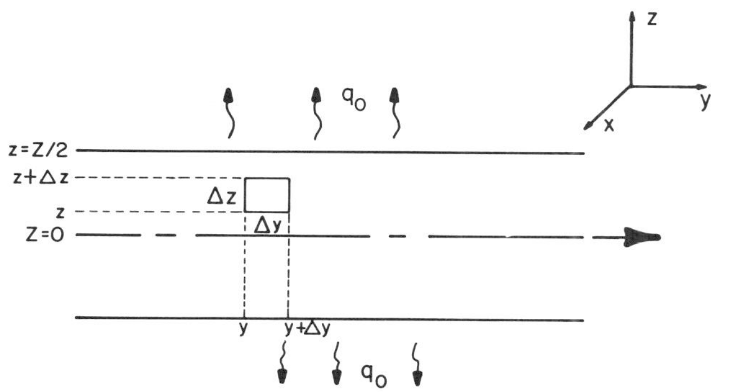

FIG. 2.5. Vertical section through a slab showing the volume element established for the heat balance.

with a unit length in the transverse (x) direction. The volume element is fixed in space, for the time being at least, so that the slab moves through it at velocity u. This is sometimes called the Eulerian coordinate framework.

A heat balance is performed over the volume element, which, as shown earlier, will yield a differential equation relating slab temperature to two dimensions, y and z. Before proceeding with the heat balance, the mechanisms of heat flow should be considered. It is readily seen that these are conduction and bulk motion (convection) in the y direction and conduction in the z direction. The terms in the heat balance can be expressed as follows:

$$\text{rate of heat input} = [\dot{q}_y|_y + u_y\rho H|_y]\Delta z \cdot 1 + \dot{q}_z|_z \cdot \Delta y \cdot 1$$

where $\dot{q}_y$ and $\dot{q}_z$ are the conductive heat fluxes in the y and z directions, respectively, and

H is the local enthalpy content of the steel per unit mass.

$$\text{rate of heat output} = [\dot{q}_y|_{y+\Delta y} + u_y\rho H|_{y+\Delta y}]\Delta z \cdot 1 + \dot{q}_z|_{z+\Delta z} \cdot \Delta y \cdot 1$$

$$\text{rate of heat generation} = \dot{q}_g \Delta y \Delta z \cdot 1$$

$$\text{rate of heat consumption} = \dot{q}_c \Delta y \Delta z \cdot 1$$

$$\text{rate of heat accumulation} = \frac{\rho(H|_{t+\Delta t} - H|_t)\Delta y \Delta z \cdot 1}{\Delta t}$$

Assembling the terms in the heat balance,

$$\begin{aligned}
&[\dot{q}_y|_y - \dot{q}_y|_{y+\Delta y} + u_y\rho H|_y - u_y\rho H|_{y+\Delta y}]\Delta z \cdot 1 \\
&\quad + [\dot{q}_z|_z - \dot{q}_z|_{z+\Delta z}] \cdot \Delta y \cdot 1 + \dot{q}_g \Delta y \Delta z \cdot 1 \\
&\quad - \dot{q}_c \Delta y \Delta z \cdot 1 = \frac{\rho(H|_{t+t} - H|_t)}{\Delta t} \Delta y \Delta z \cdot 1
\end{aligned}$$

Divide by $\Delta y \Delta z$:

$$\frac{\dot{q}_y|_y - \dot{q}_y|_{y+\Delta y}}{\Delta y} + \frac{u_y \rho H|_y - u_y \rho H|_{y+\Delta y}}{\Delta y} + \frac{\dot{q}_z|_z - \dot{q}_z|_{z+\Delta z}}{\Delta z} + \dot{q}_g - \dot{q}_c = \frac{\rho(H|_{t+\Delta t} - H|_t)}{\Delta t}$$

Now, taking the limits

$$\Delta y \to 0, \qquad \Delta z \to 0, \qquad \Delta t \to 0$$

$$-\frac{\partial \dot{q}_y}{\partial y} - \frac{\partial (u_y \rho H)}{\partial y} - \frac{\partial \dot{q}_z}{\partial z} + \dot{q}_g - \dot{q}_c = \frac{\rho \partial H}{\partial t} \tag{80}$$

Note that H is the enthalpy supplied to the volume element, or simply the heat at constant pressure

$$H - H_r^0 = \int_{T_r}^{T} C_p \, dT \tag{81}$$

Therefore,

$$\frac{\partial H}{\partial y} = \frac{\partial}{\partial T}\left[H_r^0 + \int_{T_r}^{T} C_p \, dT\right]\frac{\partial T}{\partial y} \tag{82}$$

$$\frac{\partial H}{\partial y} = C_p \frac{\partial T}{\partial y} \tag{83}$$

and

$$\frac{\partial H}{\partial t} = C_p \frac{\partial T}{\partial t} \tag{84}$$

The components of conductive heat flux, related to temperature gradients by Fourier's law, are $\dot{q}_y$ and $\dot{q}_z$.

$$\dot{q}_y = -k \frac{\partial T}{\partial y} \tag{85}$$

$$\dot{q}_z = -k \frac{\partial T}{\partial z} \tag{86}$$

For an isotropic material $k_x = k_y = k_z = k$; and for the problem of the moving slab it will be assumed that in any section of the rolling operation u_y is constant; any variations in speed will be handled in the numerical procedure adopted to solve the differential equation. Also assume that ρ is constant. Substituting these relationships into the heat balance equation, Eq. (80) the following is obtained:

$$\frac{\partial}{\partial y}\left(k\frac{\partial T}{\partial y}\right)+\frac{\partial}{\partial z}\left(k\frac{\partial T}{\partial z}\right)-u_y\rho C_p\frac{\partial T}{\partial y}+\dot{q}_g-\dot{q}_c=\rho C_p\frac{\partial T}{\partial t} \tag{87}$$

At this point, Eq. (87) can be simplified further by making the following assumptions:

1. Heat conduction in the y direction is negligible owing to the small temperature gradients (except possibly in the immediate vicinity of the surface close to the rolls and scale-breaking water jets). Certainly conduction in the y direction will be much smaller than bulk motion, and the major temperature gradients will be in the through-thickness direction. Therefore the term, $\partial/\partial y(k\,\partial T/\partial y)$ may be neglected in Eq (87).
2. There are no processes occurring during rolling whereby heat can be consumed within the slab. Thus, $\dot{q}_c = 0$ in Eq. (87). This is hardly an assumption but would be so if such endothermic processes simply were neglected.
3. With respect to the fixed observer (i.e., a stationary volume element), the rolling process operates at steady state, so the term $\rho C_p\,\partial T/\partial t = 0$ in Eq. (87).

Having made these simplifications, Eq. (87) becomes

$$\frac{\partial}{\partial z}\left(k\frac{\partial T}{\partial z}\right)-u\rho C_p\frac{\partial T}{\partial y}+\dot{q}_g=0 \tag{88}$$

At this stage it frequently is convenient to recast Eq. (88) into unsteady-state form by now allowing the observer to move with the slab. This is accomplished easily by substituting the simple relationship $y = ut$ into Eq. (88). Then, after rearranging,

$$\frac{\partial}{\partial z}\left(k\frac{\partial T}{\partial z}\right)+\dot{q}_g=\rho C_p\frac{\partial T}{\partial t} \tag{89}$$

Thus, even though the process operates at steady state, the moving observer sees the slab temperature changing with time. The moving coordinate system is sometimes termed the "Lagrangian framework."

Equation (89) obviously is the equation one would find in any elementary heat-transfer textbook for one-dimensional, unsteady-state heat conduction with a heat source term. The reader may pause to wonder why such an equation could not have been used directly without all the bother of a formal derivation. However, it should be reemphasized that Eq. (89) is valid only if all the assumptions made earlier are valid. It is vital that these assumptions are known explicitly rather than being "buried" in an "off-the-shelf" equation.

Next, in the mathematical formulation of the heat flow in rolling, the initial and boundary conditions must be specified. The initial condition can be obtained from the temperature distribution through the slab as it exits the reheating furnace. The slab temperature is likely to be uniform at this stage so that

$$t = 0, \qquad T = T_0 \tag{90}$$

From Eq. (89), which has one second-order space derivative, it is obvious that two boundary conditions are required. It is equally obvious that one of these boundary conditions should pertain to one of the slab surfaces, say, the top surface. The most useful boundary condition is the Neumann type, in which the continuity of heat flow across the boundary is expressed in terms of a heat flux, which itself may be temperature dependent, that is,

$$z = \frac{Z}{2}, \qquad -k\left.\frac{\partial T}{\partial z}\right|_{z=Z/2} = \dot{q}_0 \tag{91}$$

The external heat-flux expression will change depending on the heat-flow zone in which the slab is located. In the radiation cooling zones between rolling stands,

$$\dot{q}_0 = \sigma(T^4_{z=Z/2} - T^4_{\mathrm{a}}) \tag{92}$$

while in the descaling water jets

$$\dot{q}_0 = h(T_{z=Z/2} - T_{\mathrm{w}}) \tag{93}$$

where h is the water-jet, heat-transfer coefficient, which must be characterized empirically. Heat flow from the slab surface to the rolls may be determined from a consideration of heat conduction between semi-infinite bodies having the same temperature at the contact point, as suggested by Hollander.[21] The semi-infinite assumption is reasonable because a given area of the slab surface is in contact with the roll for only a brief time. Under these conditions,

$$\dot{q}_0 = \frac{k(T_{z=Z/2} - T_{\mathrm{c}})}{\sqrt{\pi \alpha_{\mathrm{s}} t}} \tag{94}$$

where T_{c} is the contact temperature, and
α_{s} is the thermal diffusivity.

Note that to characterize T_c, heat conduction into the roll must be considered, as described by Hollander.[21] The contact time between the slab and roll can be calculated from the reduction of the slab, the roll diameter, and the roll speed.

Having specified the heat fluxes from the surface, it remains to define the location and length of the individual cooling zones and, from the local slab velocity, the start and duration of each zone. This information is obtained from the layout of the rolling mill and the rolling schedule. Thus, as the slab moves through the rolling process the expression for $\dot{q}_0$ is selected according to the zone at any instant, and it is seen that the roll mill design and rolling schedule are incorporated into the mathematical model via the surface boundary condition.

If the heat flux in any particular zone is the same, or nearly so, from the top and bottom surfaces of the slab, the temperature distribution through it will be symmetrical with a maximum at the centerline. Then the other boundary condition can be written as follows

$$z = 0, \qquad \left.\frac{\partial T}{\partial z}\right|_{z=0} = 0 \tag{95}$$

or

$$-k\left.\frac{\partial T}{\partial z}\right|_{z=0} = 0$$

The latter expression simply states that heat does not flow across the centerline; it is effectively adiabatic. However, if heat transfer from the top and bottom surfaces is dissimilar, as determined by inplant measurement, the through-thickness temperature distribution is not symmetrical; and the second boundary condition must be specified for the bottom surface of the slab by equations such as Eqs. (92)–(94).

Equation (89) is solved subject to the initial condition, Eq. (90), and the boundary conditions, Eqs. (91) and (95), to yield the transient temperature distribution through the slab as it passes through the various zones in the rolling mill. It is clear from the relative complexity of the problem that an applicable analytical solution is unlikely to be found, and therefore a numerical solution must be sought.

Because Eq. (89) is a partial differential equation, the finite-difference method can be applied to achieve a numerical solution. Either an explicit or an implicit formulation would be suitable, although the implicit method has the advantage of stability, whereas the explicit method is subject to restrictions on the relative size of the space and time intervals. In either case, the slab is subdivided into layers of thickness, Δz, and time is discretized into intervals of Δt, as shown in Figure 2.3.

If the implicit finite-difference method is chosen, the heat-balance equation, Eq. (89), is approximated by algebraic equations derived in Section 2.3.3.F. The algebraic equations—one for each node—are solved simultaneously to yield the temperature at each node at time step $n + 1$ and at successive time steps as the calculation marches through time, commencing, of course, with the initial temperature distribution.

Up to this point nothing has been said about characterization of the slab reduction on each roll pass and the resulting generation of heat within the slab. These important phenomena have to be built into the model by an appropriate method. For example, Hollander[21] has formulated a procedure in which the shape of the through-thickness temperature profile just before a given roll pass is initially assigned to the slab after the reduction. Heat transfer to the rolls is then calculated, and the heat lost is deducted from the surface nodes only, owing to the short time of roll–slab contact. The heat generated by the rolling power is distributed uniformly to all nodes so that the resulting temperature rise is the same throughout the slab.

The mathematical model, once coded, can be run on the computer to determine the effect of the many rolling variables on the temperature distribution in the slab. However, before these predictions can be believed and used with confidence, measurements of, say, the surface temperature of the slab must be made at various locations in the rolling process and compared to the model output. This validation procedure may have to be repeated many times for a variety of rolling conditions, as heat-transfer coefficients or other parameters are fine-tuned to ensure accurate model predictions. It may be necessary to measure heat transfer to the rolls by inserting thermocouples close to the roll face and monitoring the temperature response.

2.4.2. Decarburization Rate of a Stainless Steel Heat in an AOD Furnace

The AOD (argon–oxygen decarburization) process is a relatively new innovation for the production of stainless and other types of steels. The AOD furnace is a pear-shaped vessel that is refractory lined and is equipped with several horizontally oriented tuyeres located near the bottom of the furnace, as shown in Figure 2.6. Argon and oxygen are injected through the tuyeres at a

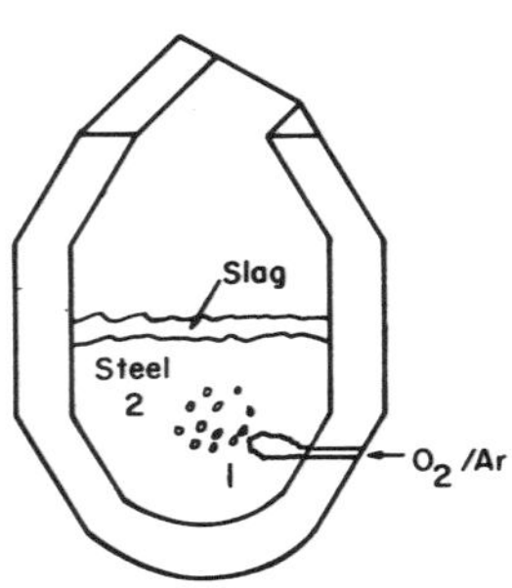

FIG. 2.6. Schematic drawing of an AOD steel-making vessel.

predetermined ratio into the steel bath to effect controlled refining via oxidation reactions.

In the production of stainless steel, the oxygen reacts with several elements dissolved in the molten steel (e.g., carbon, chromium, and silicon). Close to the tuyeres where oxygen gas contacts the liquid steel (see 1 in Figure 2.6), the chromium is oxidized as follows:

$$3O_2 + 4[Cr] = 2Cr_2O_3 \tag{96}$$

The Cr_2O_3 that is produced would eventually enter the slag phase, representing an expensive alloy loss, but the carbon dissolved in the steel reduces most of it back to Cr in regions of the bath remote from the tuyeres (see 2 in Fig. 2.6):

$$Cr_2O_3 + 3C = 2Cr + 3CO \tag{97}$$

The role of the argon that is injected with the O_2 is to lower the partial pressure of CO and thereby drive the reaction further toward products according to LeChatelier's Principle.

From the standpoint of establishing optimum AOD injection practices (e.g. O_2/Ar ratios at different stages of the blow), it is desirable to understand the influence of these variables on at least two key aspects of the process: the decarburization rate, which affects productivity, and the loss of chromium to the slag, which imposes a financial penalty on the process. To achieve this knowledge, a mathematical model based on mass transport and chemical considerations will be developed. The model was first formulated by Fruehan.[22]

Based on the conception of the AOD process described here, the following assumptions are made:

1. Most of the oxygen injected into the bath is consumed in the immediate vicinity of the tuyeres to form Cr_2O_3 (some oxygen will react with silicon in the early stages of the blow, but this is neglected here).
2. Subsequently most of the Cr_2O_3 is reduced by carbon dissolved in the steel.
3. At low carbon concentrations the rate of the Cr_2O_3 reduction reaction is limited by mass transfer of carbon to the melt-(Cr_2O_3)–bubble interface from the bulk of the bath. This has been determined in laboratory experiments. At higher carbon levels, however, all oxygen entering the bath is assumed to react directly with carbon at the tuyeres $C + 1/2O_2 = CO$.
4. Owing to the agitation introduced by the injected gases, the steel bath is backmixed.
5. The bath is at a fixed temperature (i.e., heat flow effects are ignored).

Because there are assumed to be no concentration gradients in the bath, the volume element chosen for the mathematical model is the overall volume of the bath. Then, since a key objective is to predict the decarburization rate, a molar balance on carbon is performed over the bath volume. The terms in the balance are as follows:

Rate of carbon input = 0

(no carbon enters the bath after the vessel has been charged)

Rate of carbon output = 0 (no carbon leaves the bath *as carbon*)

Rate of carbon consumption within the bath = $n_C \cdot A_v V_b$

(due to reaction with Cr_2O_3)

where n_C is the molar flux of carbon to the bubble–melt interface from the bulk, A_v is the area of the reacting interface per unit volume of bath, and V_b is the volume of the bath.

Rate of carbon generation = 0

$$\text{Rate of carbon accumulation} = V_b \frac{dC_C^b}{dt}$$

where C_C^b is the molar concentration of carbon in the bulk of the bath. Assembling the terms in the molar balance,

$$-n_C A_v = \frac{dC_C^b}{dt} \tag{98}$$

The molar flux of carbon can be expressed in terms of an empirical mass transfer coefficient and a concentration driving force:

$$n_C = k_C(C_C^b - C_C^*) \tag{99}$$

where C_C^* is the concentration of carbon at the reaction interface. Then Eq. (98) can be rewritten as

$$-k_C A_v(C_C^b - C_C^*) = \frac{dC_C^b}{dt}$$

or

$$-k_C A_v[(\%C)^b - (\%C)^*] = \frac{d(\%C)^b}{dt} \tag{100}$$

where $(\%C)^b$ is the concentration of carbon in weight percent. The product $k_C A_v$ has to be determined by experiment on an operating AOD vessel.

The reaction between Cr_2O_3 and carbon at steel-making temperatures may be considered instantaneous (recall that carbon transport is rate limiting), with the result that local equilibrium should exist at the reaction interface. Then the concentrations of reactants and products at the interface are given by the equilibrium constant K with due account taken of activity coefficient γ_C and conversion factors from mole fraction to weight percent. The concentration of carbon at the reaction interface then can be expressed as

$$(\%C)^* = \frac{a_{Cr}^{2/3} 10^2 M_C P_{CO}}{K^{1/3} M_{Fe} \gamma_C} \tag{101}$$

where K and γ_C are determined from thermodynamic data.

The partial pressure of CO can be calculated from the molar rates of argon (N_A) and CO, (N_{CO}), and the average pressure in the bath (P):

$$P_{CO} = \left(\frac{N_{CO}}{N_{CO} + N_{Ar}} \right) \cdot P \tag{102}$$

where from stoichiometry

$$N_{CO} = -\frac{10^{-2} W}{M_C} \frac{d(\%C)^b}{dt} \tag{103}$$

W is the bath mass and M_C is the atomic weight of carbon. Substituting Eqs. (101)–(103) into Eq. (100), the following expression is obtained:

$$\frac{d(\%C)^b}{dt} = k_C A_v \left[(\%C)^b + \frac{(a_{Cr}^{2/3} WP/K^{1/3} M_{Fe} \gamma_C)(d(\%C)^b/dt)}{N_{Ar} - (10^{-2} W/M_C)(d(\%C)^b/dt)} \right] \tag{104}$$

This nonlinear ordinary differential equation can be solved numerically to give the carbon concentration as a function of time for different chromium contents and argon flow rates, for example.

Equation (104) applies to the blowing period in which the carbon content is low and carbon transport is rate limiting. Earlier in the blow when the carbon level is higher and all oxygen injected reacts with carbon at the tuyeres, the decarburization rate is determined simply by stoichiometry:

$$\frac{d(\%C)^b}{dt} = \frac{-2 M_C N_{O_2}}{10^{-2} W} \tag{105}$$

where N_{O_2} is the molar rate of oxygen injection. The choice of which equation to use—Eq. (104) or (105)—is dictated by whichever predicts the lowest, that is, the limiting, decarburization rate.

The loss of chromium to the slag is the difference between the chromium oxidized at the tuyeres and the Cr_2O_3 reduced by the carbon in the bath. The oxidation rate is calculated from stoichiometry

$$(N_{Cr})_{ox} = 4/3 f N_{O_2} t \tag{106}$$

where f is an efficiency factor for chromium oxidation. The reduction rate of chromium is again determined from stoichiometry:

$$(N_{Cr})_{red} = \frac{2}{3}\left[10^{-2}\frac{W}{M_C}\{(\%C)^b - (\%C)_o^b\}\right] \tag{107}$$

where $(\%C)_o^b$ is the carbon level at which chromium oxidation at the tuyere commences. Thus, the chromium loss to the slag, $\Delta\%Cr$, is

$$\Delta\%Cr = \frac{M_{Cr}}{10^{-2}W}\left[\frac{4}{3} f N_{O_2} t - \frac{2/3(10^{-2})W}{M_C}\{(\%C)^b - (\%C)_o^b\}\right] \tag{108}$$

It should be obvious that in the early part of the blow when the carbon level is high, all (or nearly all) of the oxygen is consumed by carbon. Thus, the chromium loss is minimal.

This model of the AOD process still cannot be applied with confidence until the parameter $k_C A_v$ has been determined empirically. Fruehan[22] has described how the carbon concentration was measured at different times during the production of a 409-grade stainless steel. The measurements were made at a time when the carbon content was low ($\sim 0.15\%$) such that $(\%C)^*$ was very small so that Eq. (104) reduces to

$$\frac{d(\%C)^b}{dt} = -k_C A_v (\%C)^b \tag{109}$$

Integration of this equation from an initial time, at which the carbon content is $(\%C)_o^b$, yields

$$\ln\left[\frac{(\%C)^b}{(\%C)_o^b}\right] = -k_C A_v t \tag{110}$$

Plotting the measured carbon concentrations in this form gives a straight line having a slope of $-k_C A_v$.

2.4.3. The Entrainment of Gases by Molten Metal Streams

Figure 2.7 shows a sketch of a molten metal stream teemed from a ladle into a mold; similar situations may arise when metal streams are poured into tundishes in continuous casting and/or when metal streams are fed from a tundish into the

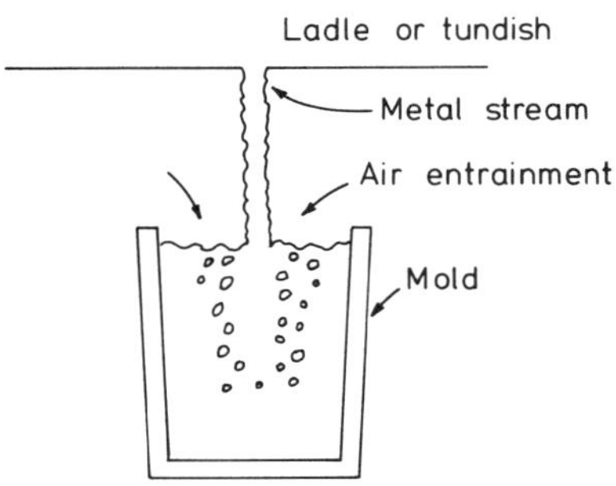

FIG. 2.7. Entrainment of air by a molten metal stream during teeming.

mold of a continuous-casting machine. The moving metal stream entrains some of the surrounding air, and the oxygen content of this entrained air may react with the stream to cause reoxidation, which would be clearly undesirable.

As discussed by Chohm et al.,[23] the actual reoxidation process may be quite complex, because under certain conditions breakup of the stream may occur, and not all the oxygen entrained may be actually absorbed by the metal pool; furthermore, the disturbance caused by the stream at the pool surface may enhance the interaction of the pool with the environment, causing additional oxygen transfer. However, one may obtain a preliminary estimate of the air entrained, and hence the oxygen transferred into the metal, by considering that the moving metal stream may be regarded as a solid rod, upon which an axisymmetric boundary layer is formed, as sketched in Figure 2.8. Then the problem is reduced to describing the formation of an axisymmetric boundary layer on a moving cylindrical rod.

If for the moment the gas is assumed to be isothermal (at some mean temperature) and incompressible, then the statement of the problem is readily given, with the aid of Table 2.2. Thus,

$$\frac{\partial}{\partial r}(ru_r) + \frac{\partial u_z}{\partial z} = 0 \quad \text{(equation of continuity)} \tag{111}$$

$$\rho\left[u_r\frac{\partial u_z}{\partial r} + u_z\frac{\partial u_z}{\partial z}\right] = \mu\left[\frac{1}{r}\frac{\partial}{\partial r}\left(r\frac{\partial u_z}{\partial r}\right) + \frac{\partial^2 u_z}{\partial z^2}\right] \quad \text{(equation of motion)} \tag{112}$$

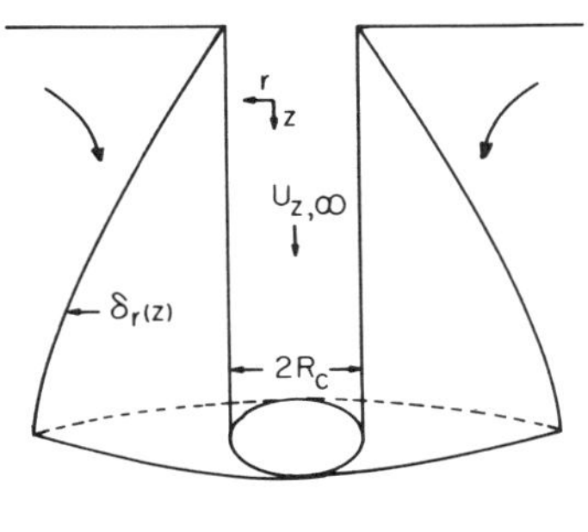

FIG. 2.8. Idealized physical picture upon which the entrainment calculation is based. The metal stream is considered to be a solid rod of radius R moving at velocity U. Also shown is the thickness of the boundary layer formed on the surface of the rod.

Note that in the statement of Eqs. (111) and (112) all the terms containing θ have been omitted; furthermore, in line with the usual boundary layer assumptions, only the z component of the equation of motion (i.e., the component in the principal flow direction) has been retained. However, both u_z and u_r were retained in the statement of the equation of continuity.

The boundary conditions for Eqs. (111) and (112) must express the following:

$$u_z = u_{z,\infty}, \qquad r = R_c \tag{113}$$

$$u_r = 0, \qquad r = R_c \tag{114}$$

$$u_z = u_r = 0, \qquad r \to \infty \tag{115, 116}$$

Here Eqs. (113) and (114) represent the so-called no-slip boundary condition, expressing the fact that the velocity of the fluid in contact with a solid surface is that of the solid surface. Furthermore, Eqs. (115) and (116) express the fact that at a long distance from the moving cylinder (more precisely, at the outer edge of the boundary layer) the velocity is zero.

The system of Eqs. (111)–(116) has been solved by Glauert and Lighthill.[24] For the present purpose, our principal interest is the extent of entrainment, which may be conveniently expressed in terms of Δ, which is the cross-sectional area equivalent to the displacement thickness, defined as

$$\Delta(z) = \int_{R_c}^{\infty} \left(1 - \frac{u_z}{u_{z,\infty}}\right) 2\pi r \, dr \tag{117}$$

Here

$$\Delta = \pi[(\delta_r^* + R_c)^2 - R_c^2] \tag{118}$$

where δ_r^* is the linear displacement thickness.† W, the mass rate of air entrainment is then given by

$$W = u_{z,\infty} \bar{\rho} \Delta \tag{119}$$

where $\bar{\rho}$ is the (mean) density of the air.

Figure 2.9 shows a plot of $\Delta/\pi R_c^2$ against $\nu z/u_{z,\infty} R_c^2$ as proposed by Glauert and Lighthill, where $\nu = \mu/\rho$ is the kinematic viscosity.[24] The use of Fig. 2.9 in conjunction with Eq. (119) provides a convenient way of *estimating* the entrainment rate. Table 2.3 shows a comparison between the predictions for the oxygen pickup by a steel stream (designated ΔO in ppm), experimental measurements by Kenny,[25] and data obtained using a more sophisticated model by Chohm et al.[23]. It is seen that the agreement between these authors is good.

†For example, see J. Szekely,[8] p. 105.

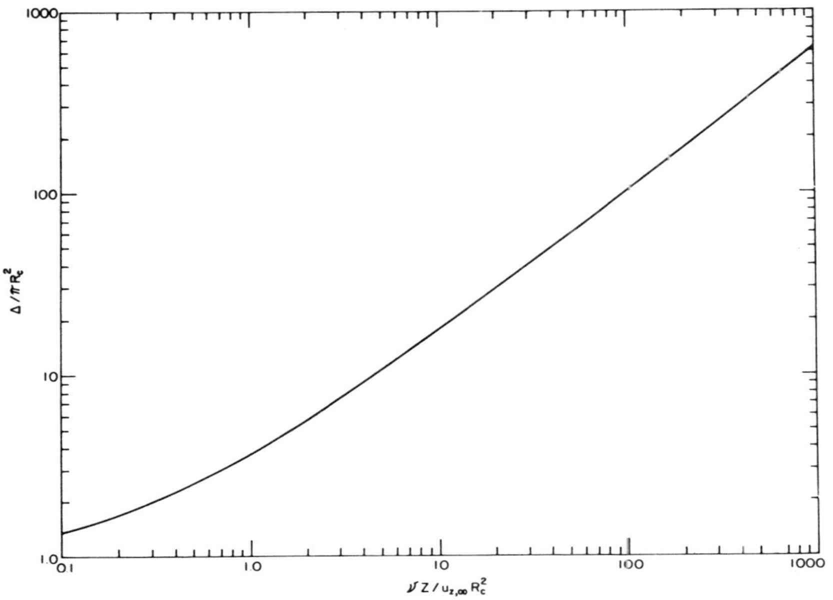

FIG. 2.9. Displacement thickness of the boundary layer formed on the thin cylindrical body. (From Szekely.[8])

TABLE 2.3. Comparison of Oxygen Pickup Calculated Using the Simple Entrainment Model Presented Here with Experimental Data by Kenny[25] and a More Complex Model Proposed by Chohm et al.[23]

Data	Kenny	Szekely (Simple Entrainment Model)	Chohm et al.
z, stream height (mm)	380	380	380
R_c, stream radius (mm)	7.9	7.9	9.0
u_z, stream velocity (m/s)	2.5 (estimated)	2.42	2.72
ΔO (ppm)	30	26.5	29.8

2.5. CONCLUDING REMARKS

This chapter has set out the principles of mathematical modeling. The mathematical model has a variety of important roles in process development, ranging from improved conceptual understanding to design and control. For the purpose of process development, mathematical modeling invariably requires concomitant measurements on the process or in the laboratory. An experimental program is necessary to provide missing information on phenomena being modeled and also to generate data for comparison to model predictions.

The development of any mathematical model follows basically the same steps from initial planning and outlining of objectives to mathematical formulation, solution of equations, model validation, and application. Of these, the mathematical formulation, which is based on the conservation principle for the mechanistic models under consideration, and solution of equations frequently consume less time than the other steps, particularly when plant trials are required.

Numerical methods are usually adopted to solve the equations resulting from applying the conservation law in the mathematical formulation stage of model development. These techniques are powerful but are necessarily approximate because they are based on the discretization of space and/or time. One of the most commonly used numerical techniques is the finite-difference method, although the finite-element method is widely used for problems involving complex geometries or stress generation.

To illustrate these principles, this chapter has given examples of the development of mathematical models for prediction of the temperature distribution in a steel slab during hot rolling, the decarburization rate and chromium loss in the AOD process, and the entrainment of air by a metal stream plunging into a bath.

REFERENCES

1. D. M. Himmelblau and K. B. Bischoff, *Process Analysis and Simulation: Deterministic Systems*, Wiley, New York, 1968.
2. N. J. Themelis, *Met. Trans.*, **3**, 2021–2029 (1972).
3. V. Venkateswaran and J. K. Brimacombe, *Met. Trans.*, **8B**, 387–398 (1972).
4. T. R. Meadowcroft and J. K. Brimacombe, *Proceedings of the Darken Conference on Physical Chemistry in Metallurgy*, U. S. Steel Corp., Monroeville, PA, 1976, pp. 121–145.
5. R. B. Bird, W. E. Stewart, and E. N. Lightfoot, *Transport Phenomena*, Wiley, New York, 1960.
6. J. Szekely and N. J. Themelis, *Rate Phenomena in Process Metallurgy*, Wiley-Interscience, New York, 1971.
7. J. Szekely, J. W. Evans, and H. Y. Sohn, *Gas–Solid Reactions*, Academic Press, New York, 1976.
8. J. Szekely, *Fluid Flow Phenomena in Metals Processing*, Academic Press, New York, 1979.
9. F. Kreith and W. Z. Black, *Basic Heat Transfer*, Harper & Row, New York, 1980.
10. S. Ostrach, *Ann. Rev. Fluid Mech.*, **14**, 313 (1982).
11. G. Oreper and J. Szekely, *J. Fluid Mech.*, **147**, 53 (1984).
12. J. Szekely and N. El-Kaddah, *Proceedings of the International Conference on Refining and Alloying of Aluminum and Ferroalloys*. Aluminum Verlag, Dusseldorf, Germany, 1985, pp. 247–266.
13. M. Abramowitz and I. A. Stegun, Eds. *Handbook of Material Findings*, National Bureau of Standards, Washington, D.C., 1964.
14. B. Carnahan, H. A. Luther, and J. O. Wilkes, *Applied Numerical Methods*, Wiley, Toronto, 1979.
15. W. F. Ames, *Numerical Methods for Partial Differential Equations*, Academic Press, New York, 1977.

16. L. W. Johnson and R. D. Riess, *Numerical Analysis*, Addison-Wesley, 1982.
17. J. R. Rice, *Numerical Methods, Software and Analysis*, McGraw-Hill, New York, 1983.
18. O. Zienkiewics, *The Finite Element Method in Engineering Science*, McGraw-Hill, London, 1971.
19. C. S. Desai and J. F. Abel, *Introduction to the Finite Element Method*, Van Nostrand Reinhold, New York, 1972.
20. K. H. Huebner, *The Finite Element Method for Engineers*, Wiley, New York, 1975.
21. F. Hollander, *Proceedings of Conference on Mathematical Models in Metallurgical Process Development*, Iron and Steel Institute, London, 1970, pp. 46–74.
22. R. J. Fruehan, *Ironmaking and Steelmaking*, **3**, 3, 153–158 (1974).
23. T. Chohm, K. Iwata, and M. Inouye, *Trans. ISI Japan*, **23**, 598 (1983).
24. M. B. Glauert and M. J. Lighthill, *Proc. R. Soc. London Ser. A*, **204**, 188 (1954).
25. M. P. Kenny, *J. Met.*, **20**, 68 (1968).

3 THE PHYSICAL MODELING OF METALS PROCESSING OPERATIONS

Chapter 2 introduced the concepts of mathematical modeling and the mathematical techniques available for the development of mathematical models. The discussion illustrated the tremendous usefulness of mathematical models for identifying relationships between key process variables and thus for the optimization of existing processes and the quantitative assessment of new process technologies. This discussion made it clear that, in order to construct a realistic mathematical model based on the fundamentals of transport phenomena (and kinetics), it is necessary to have a good physical picture of the system (e.g., we may be concerned with the flow of a gas through a packed bed, a fluidized bed operation, interaction between gas bubbles and melts, or natural convection-driven flows in melts). Then the mathematical model may be constructed by combining existing known relationships in order to represent the system.

In many cases these relationships exist (e.g., the rising velocity of gas bubbles in melts, the description of an impinging or a submerged jet, or radiation heat transfer between surfaces) and, as illustrated in Chapter 2, one may readily proceed with the development of an appropriate mathematical model. However, in many other instances one just does not have an adequate physical basis for the construction of a rigorous or even an approximate mathematical model. Furthermore, the very nature of the real industrial system may preclude direct observations from which such a physical picture may be deduced (e.g., it is difficult to observe the hearth region of the iron blast furnace, the interior of an argon stirred ladle, or the metal flow inside an aluminum cell. In yet other cases, while one could represent the system rigorously in terms of the appropriate differential equations, the system of equations that would ultimately result would be too complex or too expensive to solve.

Under such conditions we have another very useful tool at our disposal: the construction of a physical model. The description of physical modeling techniques is the subject matter of this chapter.

3.1. DEFINITION AND CLASSIFICATION OF PHYSICAL MODELS

In constructing a physical model of a process or of some particular aspect of a process, we seek to represent or model the system by changing the materials that are handled and also frequently the scale of the operations. The main objective of modeling is to achieve a realistic representation of the system using materials and equipment such that measurements may be made more conveniently and in a cost-effective manner. The key phrases in the above statement are *realistic representation* and *cost-effective manner*. The actual criteria for realistic representation, more properly stated, the *similarity criteria*, are discussed and quantified later in this chapter.

Typical examples of physical modeling include the modeling of argon-stirred ladles holding molten steel using the air–water system, the use of water in plexiglas containers to represent the behavior of ladle–tundish systems in continuous casting, the modeling of the flow of hot combustion gases using water, the use of mercury to model liquid metals in electromagnetically driven flow systems, and the like.[1–3]

In such cases, molten steel or hot combustion gases are modeled using water at room temperature so that the velocity measurements or the flow visualization studies can be readily made. Furthermore, the reduction in linear scale allows these measurements to be made cost effectively.

Physical models may be classified into the following three main groups:

1. Rigorous physical models are constructed by following certain strict rules, or similarity criteria. Under these conditions the quantitative measurements made in the course of model experiments (e.g., velocity fields, temperature profiles, and reaction rates) may be translated directly, through appropriate scaling, to describe the behavior of the real systems.
2. An alternative approach is to study the relevant physical phenomena with the objective of providing information for the construction of a mathematical model. The behavior of the real industrial system would then be predicted from the physical model.
3. A third approach is to conduct ad hoc experiments in order to acquire a feel for the system as a preliminary to mathematical modeling studies.

Each of these approaches has a definite role in research and development studies.

Rigorous physical models are helpful because they often obviate the need for parallel mathematical model development. Thus, properly scaled models have been used to deduce flue gas velocities in furnaces and to define the mixing characteristics of ladles holding molten metals, the flow fields in induction furnaces, and the like. In many practical cases, however, one simply cannot construct a model that satisfies all the similarity criteria. Under such conditions one may have to model one key aspect of the problem, and such work may have to be complemented by mathematical modeling work in any case.

The study of the relevant physical phenomena has been widely practiced and has led to the development of very important relationships for the behavior of gas bubbles in melts, the interaction of impinging jets with melts, and flow phenomena in packed bed systems.[4]

The ensuing mathematical representation of these phenomena, deduced largely from room-temperature measurements on model systems, then enables one to predict, with some confidence, the behavior of the real high-temperature systems.

Finally, ad hoc experiments aimed at flow visualization, the preliminary assessment of kinetics,[5] or the assessment of whether a system will fluidize satisfactorily, can be very useful preliminaries to a more serious research program. However, even in the design of these ad hoc measurements, care has to be taken to ensure that the data taken are representative.

In the following we shall discuss these three approaches, with emphasis on the philosophy and practice of physical model construction.

3.2. THE DEVELOPMENT OF RIGOROUS PHYSICAL MODELS

The key requirement for the model to represent the real system or the prototype is the observation of the quantitative criteria for similarity. First we shall discuss the principles of similarity, and then we shall examine the practical guidelines for ensuring that the similarity criteria are observed. There exist many states of similarity. However, here we shall confine our attention to geometric similarity, kinematic similarity, dynamic similarity, thermal similarity, and chemical similarity.

3.2.1. Geometric Similarity

Geometric similarity is the similarity of shape. Systems are geometrically similar when the ratio of any length in one system to the corresponding length in the other system is the same everywhere. This ratio is usually termed the *scale factor*. While geometric similarity is one of the most obvious requirements in modeling, often it may not be possible to attain perfect geometric similarity. Under such conditions certain critical dimensions of the model are scaled to the prototype, while other dimensions are not. This procedure is described as the use of a *distorted model*. Examples of situations in which distorted models have to be

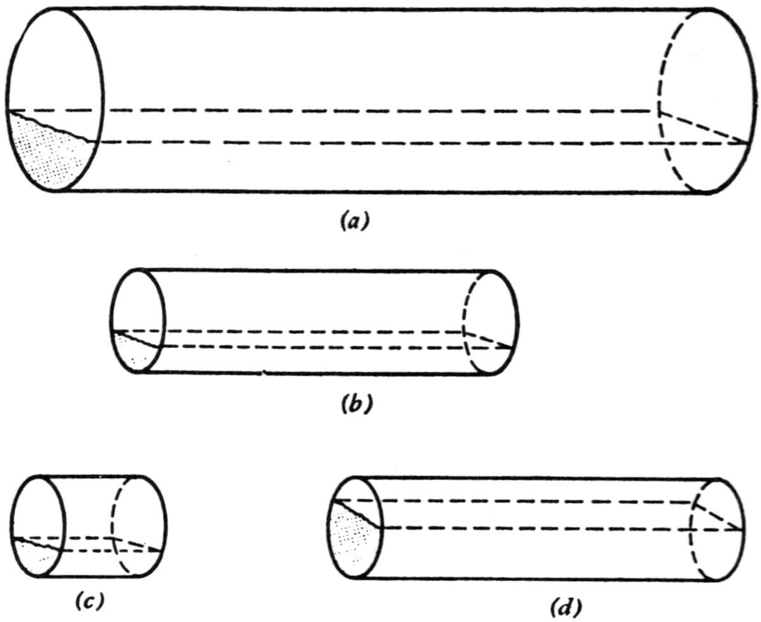

FIG. 3.1. Various types of geometric models. (*a*) Prototype. (*b*) Geometrically similar model. (*c*) Section model. (*d*) Distorted model.

used include studies in which the surface roughness may not be appropriately represented or when the proper scaling of rivers, lakes, or processing equipment would result in too shallow a liquid level.

A special case of a distorted model is the use of a section model. For example, in representing a kiln or a converter, the diameter and liquid (or solid) depth in the model may be appropriately scaled, but the model would have only a fraction of the scaled length of the prototype.

Figure 3.1 shows various types of geometric models representing the behavior of a liquid in a horizontal cylindrical reactor.

3.2.2. Kinematic Similarity

Kinematic similarity represents similarity of motion. In essence, kinematic similarity is observed between two systems if in addition to being geometrically similar, the velocities at corresponding locations in the two systems are in the same fixed ratio. Figure 3.2 illustrates the behavior of two kinematically similar systems.

3.2.3. Dynamic Similarity

Dynamic similarity represents similarity of forces. Dynamic similarity is observed between two systems when the magnitude of forces at corresponding locations in each system is in a fixed ratio. In the modeling of fluid flow

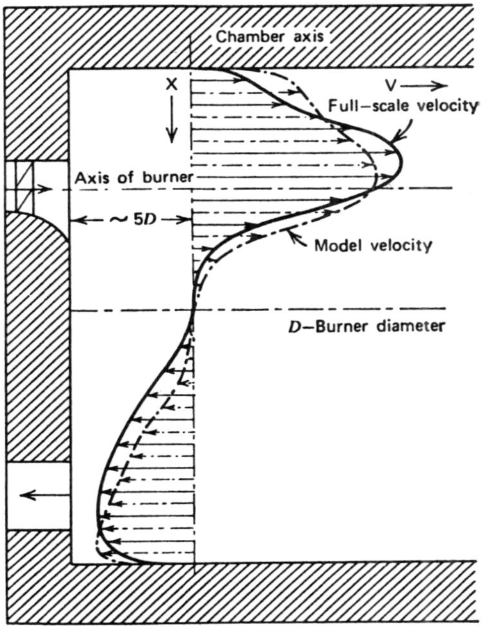

FIG. 3.2. Velocity profiles in a gas-fired heat-treating furnace and its $\frac{1}{10}$-scale water model.

problems, it is essential that both geometric and dynamic similarity be observed if quantitative deductions are to be made from the results.

3.2.4. Thermal Similarity

In nonisothermal systems, we must also satisfy the criteria for thermal similarity. In other words, the rate of heat transfer by the various mechanisms, such as conduction, convection, and radiation must be of a fixed ratio for the model and the prototype. Under such conditions, when the systems are geometrically and kinematically similar, then the temperature profiles in the model and in the prototype are in a fixed ratio throughout.

3.2.5. Chemical Similarity

Chemical similarity concerns the establishment of the necessary conditions in the model so that the rate of chemical reaction at any location is proportional to the rate of the same reaction at the corresponding time and location in the prototype. It is therefore necessary that all conditions that contribute to the overall rate of the chemical reaction be simulated in the model.

The concentration of a chemical component at a specific location in the chemical reactor or furnace depends on the initial concentration, the rate at which the particular component is produced or reacted by chemical reaction, the rate at which it diffuses (by molecular or eddy diffusion), and the rate at which it is transported by bulk movement of material to and from the location of the reaction. In turn, these phenomena depend on the temperature, the concentration gradient, and the flow pattern in the reactor. It can be concluded, therefore, that chemical similarity entails thermal and kinematic similarity, and in addition it requires a concentration profile within the model that corresponds to that of the prototype. In geometrically and chemically similar systems, the concentration profiles in the model and prototype must be geometrically similar.

In practice the chemical simulation of a process presents a number of problems because it requires the proportionality of time, temperature, and concentration between model and prototype. Usually, the rate of the chemical reaction is very sensitive to temperature, and this sensitivity limits the range of operating temperatures in the model to that of the prototype.

In many pyrometallurgical systems involving the interaction of melts (slag–metal kinetics) or gases and melts, the rates of the chemical reactions are often not rate controlling, so the attainment of chemical similarity is not an important consideration; of course, the similarity of the heat and mass transfer rates must be accomplished. In contrast, for certain gas–solid reactions, such as iron oxide reduction or coke gasification, at least partial control by chemical kinetics cannot be ruled out; here the attainment of chemical similarity is important.

In many cases it is simply impossible to satisfy all the similarity criteria simultaneously. Under such conditions, certain compromises have to be made in modeling only one key aspect of the process and then using some other means, such as mathematical modeling or complementary physical modeling techniques, to obtain a clear definition of the solution to the problem. In fact, physical modeling alone hardly ever provides the solution to a problem, because in most of the cases the interpretation of the measurements, their extrapolation, and perhaps their use in conjunction with a mathematical modeling effort is necessary.

For example, we may wish to study macrosegregation in the continuous casting of steel. Such macrosegregation may depend on the fluid flow field in the mold region, on the fluid flow field in the mushy zone, on the structure of the mushy zone, on the solidification characteristics of the system, and on the chemical nature of the impurities. Clearly, it would be impossible to model all these aspects of the problem at once. However, one may make a useful start by physically modeling the fluid flow field in the in-mold region and using this information to construct a thermal mathematical model of the system, which may then be extended to regions below the mold level.

Let us now examine the techniques available for defining the scaling factors and for the precise definition of the experimental conditions within the model to ensure its similarity to the prototype.

3.2.6. Techniques Available for Ensuring Similarity

A. Geometric Similarity

Geometric similarity of the model and the prototype is the first requirement that must be satisfied. This is readily done, for many systems, by simply keeping the ratio of any length in a system to a corresponding length in the other system the same. *In other words, we use the same scale factor for each linear dimension.*

It follows that if the prototype is a cylindrical vessel of height H and of diameter D and is filled with melt to a height H_0, then the corresponding dimensions for the prototype are βH, βD, and βH_0, where β is the scale factor. It should be noted that in most cases the quantity β cannot be selected arbitrarily but that this magnitude is defined as a result of the other constraints imposed on the system.

In some cases one cannot scale every dimension exactly; for instance, we may use a porous plug to introduce a gas into a molten metal. It may not be possible to scale the size of the pores in the porous plug in the prototype and in the model.

B. Dynamic Similarity

Since most of the modeling work involving metallurgical systems concerns fluids, the need to observe dynamic similarity is perhaps one of the most important criteria that have to be met. For this reason we shall concentrate our attention on the factors that govern dynamic similarity. Dynamic similarity represents similarity of forces. The principal forces to be considered in obtaining dynamic similarity are

inertial forces,
pressure forces,
viscous forces,
gravity,
surface tension,
elastic forces, and
electromagnetic forces.

Upon forming the ratios of some of these forces, certain key dimensionless groups or dimensionless numbers result. By ensuring that the numerical value of these dimensionless numbers is the same in the model and in the prototype, one can ascertain that dynamic similarity is observed. Let us now define these forces in terms of the principal variables.

The inertial force acting on a fluid element is the mass of the element multiplied by its acceleration. Thus, on designating proportionality by $\propto$, the characteristic length as L, the characteristic velocity as U, and the density as ρ, we may write:

$$\text{inertial force} \quad \underset{\text{(mass)}}{\propto(\rho L^3)} \times \underset{\text{(acceleration)}}{U^2/L} = \propto \rho L^2 U^2$$

The viscous force acting on the fluid may be characterized as (shear stress) × (area), that is, $\mu(du_x/dy)L^2$. On noting that $u_x \propto U$ and $y \propto L$, we have viscous force $\propto(\mu UL)$. By similar arguments, we have gravity force $\propto \rho L^3 g$, surface tension force $\propto \sigma L$, and electromagnetic (ponderomotive) force $\propto L^2(B)^2\bar{\sigma}$. From the combination of the ratios of these and other forces, the key dimensionless groups listed in Table 3.1 result.

It should be noted that the dimensionless groups listed in Table 3.1 can also be obtained from the differential equations describing these various processes. Upon making these differential equations dimensionless through the introduction of the characteristic variables, the same dimensionless groups will naturally emerge.

Because of the many forces that may occur in a given metals processing operation, there are no hard and fast rules as to which of the criteria have to be satisfied. Some useful metallurgical examples are discussed by Szekely and Themelis[6] and in references 1, 7 and 8. However, the following general comments are appropriate:

1. In modeling a system consisting of one continuous phase (e.g., a gas jet in a gas that results from forced flow), geometric similarity must be observed, and the nozzle Reynolds numbers must be the same.
2. In modeling a system consisting of one continuous phase resulting from gravity-driven flow (e.g., a teemed stream, using a submerged nozzle in continuous casting), then, in addition to geometric similarity, the Froude numbers must be the same.
3. In thermal natural convection–driven flows, geometric similarity should be observed; in addition the Grashof–Prandtl product and the Froude numbers should be the same.
4. While in principle it should be possible to model dispersed systems, such as atomization processes, the breakup of gas jets in liquids, the behavior of bubble swarms, or of slag–metal emulsions through the simultaneous adjustment of the Froude and the Weber numbers, there are many practical difficulties posed by the very high surface tension of liquid metals. For this reason most of the successful modeling efforts have addressed essentially one-phase systems.

The application of these ideas is illustrated in the following simple examples:

Example 1

Molten steel is teemed into a billet caster 0.15 × 0.15 m square through a 0.05-m diameter submerged nozzle at a mean linear velocity of 0.5 m/s. If we wish to

TABLE 3.1. Some Key Dimensionless Groups and Their Significance

Group	Formula	Nomenclature	Quantities Represented	Application
		Fluid Dynamics		
Bond number, N_{BO}	$\frac{g(\rho - \rho_f)d_v}{L\sigma}$	d_p = drop diameter ρ = drop density ρ_f = density of surrounding fluid σ = surface tension	$\frac{\text{Gravitational force}}{\text{Surface tension force}}$	Spraying
Drag coefficient, C_d	$\frac{g(\rho - \rho_f)L}{\rho\mu^2}$	L = characteristic length dimension of object ρ = density of object u = relative velocity	$\frac{\text{Gravitational force}}{\text{Inertial force}}$	Free settling
Fanning friction factor, f	$\frac{L\left(\frac{\Delta P_f}{\rho}\right)}{2u^2 1}$	L = characteristic length dimension of conduit cross section $\Delta P_f/\rho$ = friction head l = length of pipe	$\frac{\text{Shear stress}}{\text{Velocity head}}$	Pressure drop in conduits
Froude number, N_{Fr}	$\frac{u^2}{gL}$	L = characteristic length dimension of system	$\frac{\text{Inertial force}}{\text{Gravitational force}}$	Wave and surface behavior
Froude number, modified, N'_{Fr}	$\frac{\rho_g u^2}{(\rho_l - \rho_g)gL}$	ρ_g = gas density ρ_l = liquid density	$\frac{\text{Inertial force}}{\text{Gravitational force}}$	Fluid behavior of gas–liquid systems

(continued on next page)

TABLE 3.1. Continued Some Key Dimensionless Groups and Their Significance

Group	Formula	Nomenclature	Quantities Represented	Application
Galileo number, N_{Ga}	$\frac{g\rho^2 L^3}{\mu^2}$	μ = viscosity	$\frac{(\text{Inertial}) \times (\text{Gravit.})}{(\text{viscous force})^2}$	Flow in baths of viscous liquids
Mach number, N_{Ma}, M	$\frac{\mu}{u_s}$	u = fluid velocity u_s = velocity of sound in fluid	Compressible flow	
Newton inertial force group, N_I	$\frac{F}{\rho u^2 L^2}$	F = force imposed ρ = fluid density	$\frac{\text{Imposed force}}{\text{Inertial force}}$	Agitation
Morton number, N_{Mo}	$\frac{g\mu_l^4}{\rho_l (L\sigma)^3}$	μ_l = liquid viscosity	$\frac{(\text{Gravitational}) \times (\text{Viscous})}{\text{Surface tension force}}$	Velocity of bubbles in liquids
Power number, N_P	$\frac{PL}{\rho n^3 L^5}$	P = power input to agitator L = characteristic dimension of agitator paddle n = angular speed of rotation	$\frac{\text{Drag force on paddle}}{\text{Inertial force}}$	Power consumption in agitated vessels
Reynolds number, N_{Re}	$\frac{Lu\rho}{\mu}$	L = characteristic length dimension of system ρ = fluid density μ = fluid viscosity	$\frac{\text{Inertial force}}{\text{Viscous force}}$	Fluid flow
Weber number, N_{We}	$\frac{\rho L u^2}{\sigma}$	σ = surface tension	$\frac{\text{Inertial force}}{\text{Surface tension force}}$	Bubble formation, atomization of liquid jet

Heat Transfer

Biot number, N_{Bi}	$\frac{hr_m}{k}$	h = heat transfer coefficient k = thermal conductivity r_m = distance from midpoint to surface	$\frac{\text{Internal thermal resistance}}{\text{Surface film resistance}}$	Unsteady state heat conduction
Fourier number, N_{Fo}	$\frac{kt}{\rho C_p r_m^2}$	k = thermal conductivity t = time of heating C_p = specific heat of material ρ = density of material	$\frac{\text{Time elapsed}}{\text{Time required to heat solid to steady state condition}}$	Unsteady state heat conduction
Grashof number, N_{Gr}	$\frac{g\rho^2 L^3 \beta \Delta T}{\mu^2}$	L = characteristic length dimension ρ = density of fluid β = coefficient of thermal volume expansion $\beta = -\frac{1}{\rho}\left(\frac{\partial \rho}{\partial T}\right)_p$ ΔT = temperature difference	$(N_{Re})\left(\frac{\text{Buoyancy force}}{\text{Viscous force}}\right)$ $= (N_{Ga})\beta \Delta T$	Free convection
Nusselt number, N_{Nu}	$\frac{hL}{k}$	h = heat transfer coefficient L = characteristic dimension k = thermal conductivity of fluid	$\frac{\text{Total heat transfer}}{\text{Conductive heat transfer}}$	Forced convection
Peclet number, N_{Pe}	$\frac{Lu\rho C_p}{k} = \frac{Lu}{\alpha}$	u = fluid velocity C_p = specific heat α = thermal diffusivity	$\frac{\text{Heat transfer by bulk motion}}{\text{Conductive heat transfer}}$ $= N_{Re} N_{Pr}$	Forced convection

(continued on next page)

TABLE 3.1. Continued   Some Key Dimensionless Groups and Their Significance

Group	Formula	Nomenclature	Quantities Represented	Application
Prandtl number, N_{Pr}	$\dfrac{C_p \mu}{k}$	C_p = specific heat of fluid μ = viscosity k = thermal conductivity	$\dfrac{\text{Momentum diffusivity}}{\text{Thermal diffusivity}}$ $= (N_{\mathrm{Gr}})(N_{\mathrm{Pr}})$	Forced and free convection
Rayleigh number, R'	$\dfrac{g\rho^2 L^3 C_p \beta \Delta T}{k}$	β = coefficient of thermal volume expansion ΔT = temperature difference across film L = characteristic length dimension	$\dfrac{\text{Heat transfer by convection}}{\text{Heat transfer by conduction}}$	Free convection
Stanton number, N_{St}	$\dfrac{h}{\rho C_p u}$	h = heat transfer coefficient u = velocity	$\dfrac{\text{Total heat transfer}}{\text{Thermal capacity of fluid}}$ $= \dfrac{N_{\mathrm{Nu}}}{(N_{\mathrm{Re}})(N_{\mathrm{Pr}})}$	Forced convection
Thring radiation group	$\dfrac{\rho C_p u}{\varepsilon \sigma T^3}$	ε = emissivity of surface σ = Stefan–Boltzmann radiation constant T = absolute temperature of surface	$\dfrac{\text{Bulk transport of heat}}{\text{Heat transferred by radiation}}$	Radiation
		Mass Transfer		
Grashof number, N'_{Gr}	$\dfrac{g\rho^2 L^3 \beta' \Delta X}{\mu^2}$	β' = coefficient of density change with concentration $= -\dfrac{1}{\rho}\left(\dfrac{\partial \rho}{\partial X}\right)_T$ ΔX = concentration of driving force	$(N_{\mathrm{Re}})\left(\dfrac{\text{Buoyancy force}}{\text{Viscous force}}\right)$ $= N_{\mathrm{Ga}} \beta' \Delta T$	Free convection
Lewis number, N_{Le}	$\dfrac{\rho C_p D_{\mathrm{A-B}}}{k}$	ρ = density of fluid C_p = specific heat $D_{\mathrm{A-B}}$ = molecular diffusivity k = thermal conductivity	$\dfrac{\text{Molecular diffusivity}}{\text{Thermal diffusivity}}$	Combined mass and heat transfer

Peclet number (for mass transfer), N'_{Pe}	$\frac{Lu}{D'_v}$	L = characteristic length D'_v = characteristic diffusion coefficient	$\frac{\text{Bulk mass transport}}{\text{Mass transfer by diffusion}}$	Mass transfer in reactors
Schmidt number, N_{Sc}	$\frac{\mu}{\rho D_{A-B}}$	μ = fluid viscosity ρ = fluid density D_{A-B} = molecular diffusivity	$\frac{\text{Momentum diffusivity}}{\text{Molecular diffusivity}}$	Forced and free convection
Sherwood number, N_{Sh}	$\frac{k_d L}{D_{A-B}}$	k_d = mass transfer coefficient L = characteristic dimension D_{A-B} = molecular diffusivity	$\frac{\text{Total mass transfer}}{\text{Molecular diffusivity}}$	Forced and free convection
		Chemical Reaction		
Arrhenius group	$\frac{E}{RT}$	E = activation energy R = universal gas constant T = absolute temperature	$\frac{\text{Activation energy}}{\text{Potential energy of fluid}}$	Reaction rate
Damköhler group I	$\frac{r_x L}{u c_A}$	c_A = concentration of reacting species r_x = reaction rate L = characteristic length dimension u = fluid velocity	$\frac{\text{Chemical reaction rate}}{\text{Bulk mass flow rate}}$	Combined chemical reaction, momentum, and heat transfer
Damköhler group II	$\frac{r_x L^2}{D_{A-B} c_A}$	D_{A-B} = molecular diffusivity	$\frac{\text{Chemical reaction rate}}{\text{Molecular diffusion rate}}$	Combined chemical reaction, momentum, and heat transfer

(*continued on next page*)

TABLE 3.1. Continued Some Key Dimensionless Groups and Their Significance

Group	Formula	Nomenclature	Quantities Represented	Application
Damköhler group III	$\frac{Qr_x L}{C_p \rho u T}$	Q = heat generated per unit mass reacted T = temperature C_p = specific heat ρ = density	$\frac{\text{Heat generated}}{\text{Bulk transport of heat}}$	Combined chemical reaction, momentum, and heat transfer
Damköhler group IV	$\frac{Qr_x L^2}{kT}$	k = thermal conductivity	$\frac{\text{Heat generated}}{\text{Conductive heat transfer}}$	Combined chemical reaction, momentum, and heat transfer
		Electromagnetic Phenomena		
Magnetic Reynolds number	$N_{\mathrm{Re,m}} = \frac{U_0 L_0 \sigma_e}{\mu_m}$	U_0 = fluid velocity L_0 = length σ_e = electric conductivity μ_m = magnetic permeability	$\frac{\text{Magnetic convection}}{\text{Magnetic diffusion}}$	Electromagnetically driven flows
Hartmann number	$N_{\mathrm{Ha}} = \left(\frac{\sigma_e}{\mu}\right)^{\frac{1}{2}} B_0 L_0$	B_0 = magnetic flux intensity μ = fluid viscosity	$\frac{\text{Ponderomotive force}}{\text{Viscous force}}$	Electromagnetically driven flows
Magnetic Prandtl number	$N_{\mathrm{Pr,m}} = \frac{N_{\mathrm{Re}}}{N_{\mathrm{Re,m}}} = \nu \sigma_e \mu_m$	$\nu = \frac{\mu}{\rho}$	$\frac{\text{Ability of fluid to transmit magnetic field}}{\text{Ability of fluid to transmit vorticity}}$	Electromagnetically driven flows
Magnetic interaction parameter	$N_m = \frac{N_H^2}{N_{\mathrm{Re}}} = \frac{\sigma_e B^2 L}{U_0 \rho}$			Magnetic damping of convection

construct a water model for this process that is twice the size of the prototype, we must define the conditions in the model.

Solution. Since the flow is gravity driven, the only criteria to be considered are

1. The model and the prototype must be geometrically similar.
2. The Froude number must be the same for the model and the prototype.
3. Conditions must be such that the flow is turbulent in the nozzle for both the model and the prototype.

Geometric similarity is readily observed. We note here that there are many occasions when the model is larger than the prototype, especially when the prototype itself is not very big and when one wishes to make precise measurements.

Condition 2 may be written as

$$\frac{u_{\rm m}^2}{gL_{\rm m}} = \frac{u_{\rm p}^2}{gL_{\rm p}}$$

Thus,

$$\frac{u_{\rm m}}{u_{\rm p}} = \sqrt{\frac{L_{\rm m}}{L_{\rm p}}} \tag{2}$$

Thus, the linear velocity exiting the nozzle in the model has to be $u_{\rm m} = \sqrt{2} \times u_{\rm p} = 0.71$ m/s.

Regarding condition 3, in the prototype,

$$N_{\rm Re} = \frac{0.5 \times 7.1 \times 10^3 \times 5 \times 10^{-2}}{5 \times 10^{-3}} = 3.55 \times 10^4 \quad \text{(i.e., turbulent)}$$

In the model,

$$N_{\rm Re} = \frac{0.71 \times 10^3 \times 0.1}{10^{-3}} = 7.1 \times 10^4 \quad \text{(i.e., turbulent)}$$

EXAMPLE 2

In the design of a leaching plant for the processing of molybdenite concentrate, the vortex configuration in the agitated autoclaves must be estimated (Figure 3.3). In order to do so, the flow conditions in the industrial autoclave are simulated in a small flow model. What should the various scaling factors in the model be in order to produce a vortex similar to that in the industrial vessel?

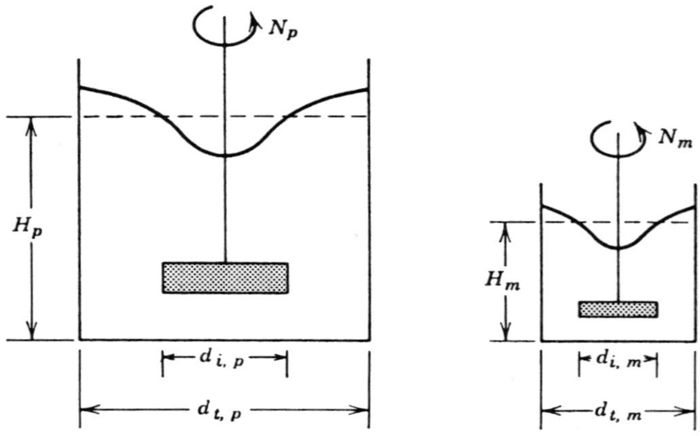

FIG. 3.3. Illustration of the prototype and its model system for an agitated vessel.

Solution. Let us select the impeller diameter and its peripheral velocity as the characteristic length and velocity in the system. The geometric similarity is observed as

$$\left(\frac{d_t}{d_i}\right)_m = \left(\frac{d_t}{d_i}\right)_p \tag{3}$$

and

$$\left(\frac{H}{d_{i,m}}\right) = \left(\frac{H}{d_{imp}}\right) \tag{4}$$

where d_t = tank diameter,
d_i = impeller diameter,
H = depth of liquid in tank at rest, and
m, p = subscripts denoting the model and the prototype, respectively.

The forces acting on the liquid are the inertia forces due to the rotation of the impeller, the shear stress due to viscosity, and the gravity force, which tends to restore the configuration that the liquid has to rest. The ratios of these forces can be expressed by the following dimensionless groups (Table 3.1):

$$\text{Reynolds number} = \frac{\text{inertial force}}{\text{viscous force}} = \left(\frac{d_i^2 N\rho}{\mu}\right)_m = \left(\frac{d_i^2 N\rho}{\mu}\right)_p \tag{5}$$

$$\text{Froude number} = \frac{\text{inertial force}}{\text{gravity force}} = \left(\frac{d_i N^2}{g}\right)_m = \left(\frac{d_i N^2}{g}\right)_p \tag{6}$$

where N = speed of impeller rotation in rev/s (thus, Nd corresponds to a velocity term),
ρ = liquid density, and
g = gravitational acceleration.

We can start the design of the model by selecting a convenient ratio for the length scale of model/prototype ($d_{i,j}/d_{i,p}$). We may then use Eq. (6) to calculate the proper impeller speed ratio to satisfy the Froude number criterion:

$$\frac{N_m}{N_p} = \left(\frac{d_{i,p}}{d_{i,m}}\right)^{1/2} \tag{7}$$

Finally, by substituting for N_m/N_p from Eq. (7) in Eq. (5) we obtain

$$\left(\frac{\mu}{\rho}\right)_m = \left(\frac{\mu}{\rho}\right)_p \left(\frac{d_{i,m}}{d_{i,p}}\right)^{3/4} \tag{8}$$

Eq. (8) shows that the only way of obtaining similar vortices in the model and the prototype is to use a model fluid of lower kinematic viscosity (μ/ρ) than in the prototype.

The consideration of force balances is just one of the techniques that one may use in establishing the criteria for dynamic similarity. The other techniques available include dimensional analysis and the use of differential equations in order to identify the key dimensionless groups.

Example 3[10]

Figure 3.4 shows a sketch of an electroslag refining unit. It is seen that this consists of a consumable electrode immersed in a molten slag phase through which a current is being passed. The passage of the current causes "joule heat generation" in the slag, which in turn causes the electrode to melt. The molten metal droplets fall through the slag, and some of the chemical impurities are removed during that stage. The metal droplets accumulate in the form of a molten metal pool, which then solidifies as an ingot.

Solution. This is clearly a complex process, but one that has been extensively modeled mathematically. In seeking to use a physical model to test the appropriateness of the mathematical model, we have to restrict ourselves to some specific aspects of the system.

Thus, while recognizing that the fluid flow field in the slag phase is driven by the combination of buoyancy and electromagnetic forces, we shall restrict our attention to the electromagnetically driven component of the flow. The circulation pattern that one would then expect in the slag is sketched in Figure 3.5. This system may be modeled by using a horizontal trough containing mercury through which a current is being passed, as shown in Figure 3.6.

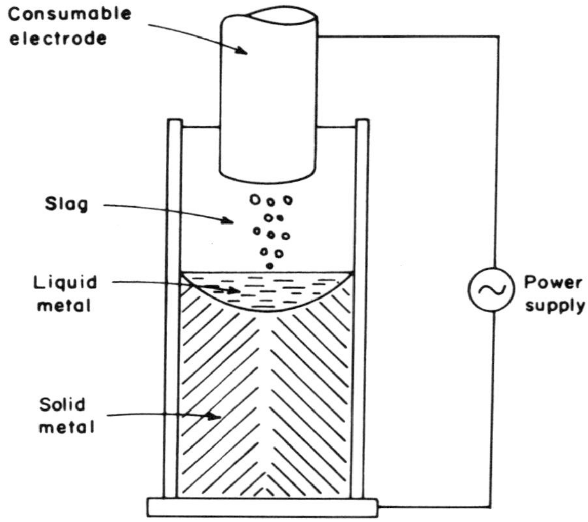

FIG. 3.4. Schematic sketch of an ESR unit.

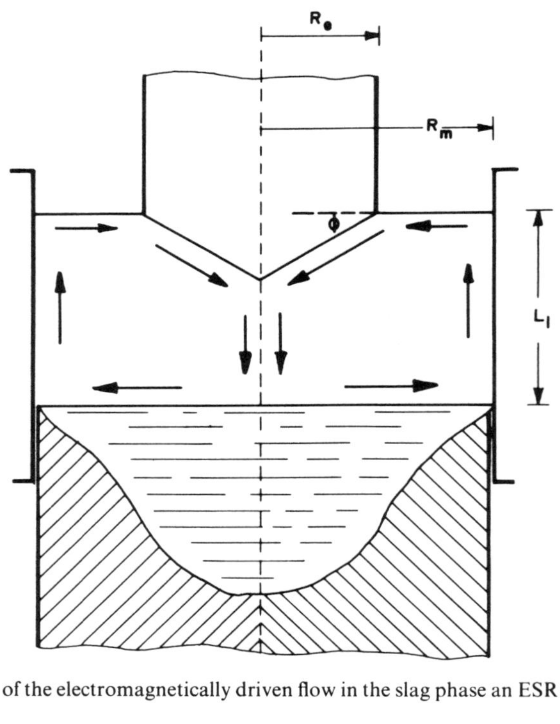

FIG. 3.5. Sketch of the electromagnetically driven flow in the slag phase an ESR unit. L is the depth of the slag pool.

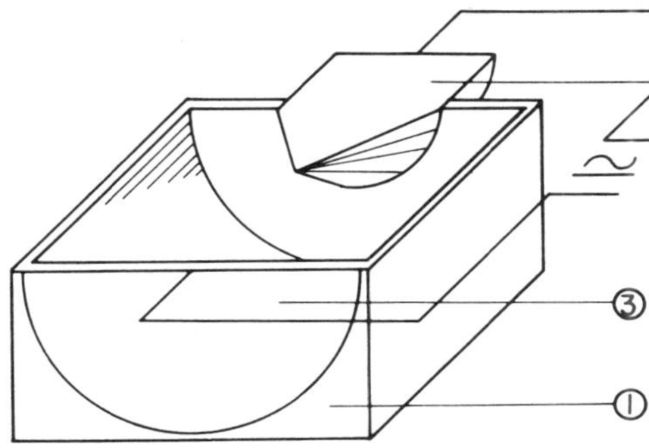

FIG. 3.6. Schematic sketch of the experimental arrangement (1, horizontally placed half-cylinder trough; 2, electrode with a conical tip; 3, copper disk representing the top surface of the metal pool).

We seek to attain geometric similarity and dynamic similarity at the plane corresponding to the free surface of the physical model, which will correspond to a vertical central section of the slag phase in the real system. Geometric similarity is readily attained. In seeking dynamic similarity with respect to the electromagnetically driven flow component, we need to consider the Reynolds number and the magnetic Reynolds number. In this case it is not necessary that the values of the magnetic Reynolds numbers be the same; it is sufficient that the magnetic Reynolds number should be much less than unity for both the model and the prototype.

A small value of the magnetic Reynolds number ensures that the rate at which the magnetic field propagates is much faster than the convective processes within the system. Thus, the effect of fluid convection on the magnetic field may be neglected. Some numerical values of the system parameters for the physical model are given in Table 3.2, while Table 3.3 provides a comparison of some key dimensionless parameters pertaining to the model and to the prototype. Since the magnetic Reynolds number, Re_m is small in both cases, we may neglect the convection of the magnetic field.

TABLE 3.2. Dimensions and Operating Conditions for the Physical Model

Parameter	Symbol	Value
Depth of mercury pool	L_1	0.13 m
Electrode diameter	$2R_e$	0.077 m, 0.108 m
Trough diameter	$2R_m$	0.140 m
Total current	I_0	0.5–2.5 kA
Angular frequency of current	ω	0–314 radians/s
Angle of electrode taper	ϕ	30°

TABLE 3.3. Dimensionless Parameters for MHD Phenomena in the Slag Region of ESR Systems

Dimensionless Parameters	Expression[a]	Value: Physical Model	Value: Industrial-Scale ESR Unit
Magnetic Reynolds number, Re_m	$LU_0\sigma_0\mu_0$	6.5×10^{-3}–3.2×10^{-2}	1×10^{-4}
Fill ratio, F	$\left(\frac{R_e}{R_m}\right)^2$	0.30–0.59	0.49
Reynolds number, Re_L	$\frac{LU_0\rho}{\mu}$	4×10^4–2×10^5	4.3×10^4

[a] U_0 is the characteristic velocity, and R_e and R_m are the radii of the electrode and the mold, respectively.

The "fill ratio" F used in the physical model falls within the range of values encountered in practice, thus satisfying the criterion of geometric similarity. Finally, most important of all, the Reynolds numbers are in the range corresponding to turbulent conditions.

The Reynolds number is a key factor in characterizing the velocity fields in the real system and in the model. It has to be stressed that the current, the density, and the magnetic permeability of the fluid enter these considerations directly, since

$$U_0 = \sqrt{\frac{\mu_0}{\rho}}\, J_0 L \tag{9}$$

It follows, that the model used is indeed representative of industrial conditions. Thus, the relationships deduced to exist between the process parameters (e.g., the relationship between the current and the linear velocity) based on the model experiments should be applicable to industrial conditions.

It is of interest to note that the use of mercury as the working fluid, with a somewhat reduced linear scale and a substantially reduced applied current, did indeed meet the similarity criteria to represent the electromagnetically driven flow component in the slag phase of an ESR unit. However, as is often the case, *we were able to model only one aspect of the problem, namely, the electromagnetically driven flow.*

Up to this point we have taken a relatively ad hoc approach to defining the key dimensionless numbers needed to establish the dynamic similarity between the model and the prototype. Alternative, more formal techniques involve the use of *dimensional analysis* or the appropriate differential equations as the starting point.

1. Dimensional Analysis. A full discussion of this technique, is available in Szekely and Themelis.[6] Here we shall confine our attention to the statement of Buckingham's π theorem, which expresses the following:

1. The solution of any dimensionally homogeneous equation representing physical quantities can be expressed in terms of a certain number of dimensionless groups, π, say,

$$f(\pi_1, \pi_2, \pi_3, \ldots) = 0 \tag{10}$$

2. If these quantities are independent, then the number of dimensionless groups needed to define the problem equals the number of individual variables minus the number of fundamental magnetudes (mass, length, time, etc.) needed to express these variables.

Let us illustrate the use of the theorem with the following simple example.

Example 4

Let us consider that the force (F) exerted on a body immersed in a moving fluid depends on the velocity (U), density (ρ), and viscosity (μ) of the fluid and on the characteristic dimension of the body L.

Solution. Using Buckingham's theorem, we must define the number of dimensionless groups required to represent the problem. The number of variables is five, and the basic dimensions are mass, length, and time or mass, length, and velocity. Thus, we need $5 - 3 = 2$ dimensionless groups to represent the problem. It may be shown that these two dimensionless groups are

$$\text{Reynolds number} = \frac{UL\rho}{\mu}$$

and

$$\text{Euler number} = \frac{F}{\rho u^2 L^2}$$

Since F/L^2 is a force per unit area, say τ_w, the shear stress $\tau_w/\rho u^2$ may also be regarded as the friction factor. Thus, the solution to our problem is

$$f\left(\frac{uL\rho}{\mu}, \frac{F}{\rho u^2 L^2}\right) = 0 \tag{11}$$

If we were to introduce another variable, say, e/L, the relative roughness of the surface, then the fundamental dimensions remain unchanged, while the number of variables is increased to six. Under these conditions we need three dimensionless groups to represent the problem:

$$f\left(\frac{uL\rho}{\mu}, \frac{\bar{F}}{\rho u^2 L^2}, \frac{e}{L}\right) = 0 \tag{12}$$

The form of Eq. (12) is, of course, consistent with the friction factor correlations available in the literature.[4] Example 4 illustrates both the strength and the weakness of the dimensional analysis approach. It was a very straightforward matter to organize the variables into familiar, dimensionless groups. However, the technique does not provide guidance as to the form of the relationship to be sought between these variables. Furthermore, dimensional analysis cannot tell us whether a key parameter or parameters have been omitted from consideration. Note that *for most of the practical problems one is likely to encounter in metals processing operations, the appropriate dimensionless groups will have been established.*

2. Differential Equations. Another technique for defining the key dimensionless groups is to develop the appropriate differential equations needed to represent the problem. Upon making these equations dimensionless, the appropriate dimensionless groups will become apparent, as illustrated in the following example.

Example 5

We must show the principal dimensionless groups that emerge when the Navier–Stokes equations are made dimensionless.

Solution. The equation of continuity is written as

$$\nabla \cdot \mathbf{U} = 0 \tag{13}$$

The equation of motion is given as

$$\frac{D\mathbf{U}}{Dt} = -\nabla P + \mu \nabla^2 \mathbf{U} + \rho \mathbf{F} \tag{14}$$

Let us define the following dimensionless parameters:

$$U^* = \frac{\mathbf{U}}{U_0}, \qquad x^* = \frac{x}{L}, \qquad y^* = \frac{y}{L}, \qquad z^* = \frac{z}{L},$$

that is $\nabla^* = L\nabla$, $(\nabla^*)^2 = L^2\nabla$, $P^* = \dfrac{P}{\rho U_0^2}$, $t^* = t\dfrac{L}{U_0}$, and so on.

Substitution of these quantities into Eqs. (13) and (14) yields, after some algebra,

$$\nabla^* \cdot U^* = 0 \tag{15}$$

and

$$\frac{DU^*}{Dt^*} = -\nabla \mathbf{P}^* + \left(\frac{\mu}{LU_0\rho}\right)\nabla^{*2}U^* + \left(\frac{gL}{U_0^2}\frac{\mathrm{F}}{g}\right) \tag{16}$$

here $LU_0\rho/\mu$ is the Reynolds number, and gL/U_0^2 is the Froude number.

The development of the key dimensionless groups through the use of a differential equation type formulation is highly desirable, particularly when considering a new system that has not been extensively studied. Under these conditions the statement of the appropriate differential equations and their associated boundary conditions forces one to examine the problem in detail and thus to obtain a checklist so that no important parameter is overlooked.

C. Thermal Similarity

At our present state of knowledge, we would not consider developing physical models for purely heat conduction problems because transient one-, two-, or even three-dimensional heat conduction is now readily tackled using numerical methods in an essentially routine manner.

Physical modeling may be usefully undertaken in cases of thermal buoyancy-driven flows, such as convection in ladles or molds. In addition to establishing geometric similarity in the study of these systems, it is also necessary that the Grashof and Prandtl numbers be the same. A problem of this type has been discussed by Hlinka.[1]

Two complicating factors should be noted here:

1. For metals the Prandtl number will be significantly less than unity, which means that the thermal boundary layer will be thicker than the velocity boundary layer. Another important consequence of this finding is that for transient systems (i.e., in heating cooling or solidifying melts held in containers) thermal equilibrium will be attained much faster than equilibrium with respect to fluid flow; in other words the fluid flow field will lag behind the temperature field. This facet of the problem is not reproduced by nonmetallic liquids.
2. In most thermal problems of practical importance one cannot specify the inner wall temperatures of a container, but rather more complex boundary conditions have to be introduced, involving transient conduction through walls of finite thickness followed by an additional thermal interchange with a heater or the environment.

There may be serious practical problems in attaining true similarity for these two sets of circumstances. However, useful physical modeling of convection in glass tanks by using glycerol or other viscous fluids has been reported.

D. Chemical Similarity

For the reasons discussed earlier, the attainment of chemical similarity between the model and a prototype would be extremely difficult. In studies involving reacting systems where the rates of chemical reaction are chemically controlled, the best way to proceed is to obtain the actual rate expressions, by separate experimentation and then incorporate these into a suitably developed mathematical model.

The predictions of the mathematical model should then be checked against measurements obtained *in a pilot plant*, which would be just a smaller scale version of the prototype. In scaling the results obtained in the pilot plant, use should be made of the mathematical model.

E. Scaling

One of the really important considerations in the construction of a physical model is to decide on a proper scale. Under certain conditions, when the requirements of similarity specify that a number of dimensionless groups be identical, the actual scale may be uniquely fixed by these considerations.[1]

Under other conditions (e.g., in many forced flow situations such as in modeling fluid flow in tundishes, argon-stirred ladles, and molds of continuous-casting units) we have to satisfy geometric similarity and dynamic similarity such that the Froude numbers are equal.

As illustrated in Example 1, under these conditions we are at liberty to select the length scale somewhat arbitrarily. As a practical consideration, we wish to observe the constraints that the flow should be turbulent in the model if this is the case for the prototype. Furthermore, one should select the linear scale such that wall effects are minimixed. In practice, people tend to work at 1:1 scale in many steel-making installations in modeling tundishes or ladles, although a reduction in linear scale by a factor of about 2:3 would probably be acceptable and offer considerable savings in cost.

Under certain conditions it is necessary to impose a distortion in the linear scale in order to avoid dominance by some extraneous factor. For example, in modeling flow and circulation in Lake Erie—which is several hundred miles long and 30–40 miles wide but only a few hundred feet deep—in observing geometric similarity either one would have to build a very large model or the pool would be so shallow that surface tension effects would become dominant. When operating with a distorted scale, great care has to be taken in the interpretation of the results and frequently recorse must be made to the use of mathematical models.

3.3. SEMIRIGOROUS MODELS: THE STUDY OF PHYSICAL PHENOMENA

As noted earlier, the development of rigorous models is often not possible because one simply cannot satisfy all the similarity criteria. For example, if we are interested in modeling the behavior of a gas bubble swarm in a molten metal pool, we need to satisfy simultaneously the equality of the Froude and the Weber numbers in the model and prototype, which would not be possible. Similarly, it may not be practical to model the atomization of molten metals using room-temperature liquids because the high surface tension of the former could not be reproduced.

It may also be very difficult to represent chemical similarity or to model electromagnetically driven flows using room temperature fluids. In the latter case, another complication may arise because, in the majority of electromagnetically driven flows, thermal energy is also dissipated in the system, giving rise to buoyancy-driven flows as well. In order to reproduce the relative importance of the Lorentz forces and the buoyancy forces, the electrical conductivities and thermal properties (e.g., the expansion coefficient of the model and the prototype) fluids would have to be appropriately scaled.

The difficulties inherent in satisfying all the similarity criteria have stimulated alternative approaches that in essence involve the identification of key phenomena within the system and the development of an experimental–analytical study of these phenomena, with the objective of obtaining a quantitative relationship between the key process variables. Such a relationship may be extrapolated under certain circumstances to the real system (i.e., the prototype). Alternatively, the experimental data thus obtained may be used to verify a mathematical model, which then can be used with some confidence to represent the industrial system. Some simple examples of this approach follow.

The pressure drop through packed beds of solids is given by the Ergun equation:

$$\frac{\Delta P}{Z} = 150\frac{(1-\varepsilon_v)^2}{\varepsilon_v}\frac{\mu U_0}{(\phi_s d_p)^2} + 1.75\frac{1-\varepsilon_v}{\varepsilon_v^2}\frac{\rho U_0}{\phi d_p} \tag{17}$$

where Δp is the pressure drop (N/m^2),
Z is the bed height (m),
ε_v is the void fraction in the bed,
μ is the viscosity of the fluid (kg/m-s),
U_0 is the fluid velocity through the empty column (m/s),
ϕ is the shape factor of the particles,
d_p is the particle diameter (m), and
ρ is the fluid density (kg/m^3).

While this equation has been developed from measurements obtained with glass beads and air at room temperature, it has been shown to apply to high-temperature systems.

It has been shown, furthermore, that the Ergun equation may be readily generalized to a multidimensional system by expressing it in vectorial form:

$$\nabla P = \mathbf{U}(f_1 + f_2 U) \tag{18}$$

where f_1 and f_2 are simply coefficients.

Another example is that the rising velocity of large ($d_p > 2$ cm) spherical cap-shaped gas bubbles in liquids is given by the relationship

$$U_b = 1.02\,(gd_B/2)^{1/2} \tag{19}$$

where U_b is the rising velocity,
g is the acceleration due to gravity, and
d_B is the bubble diameter.

It has been shown that this relationship applies equally well to molten metals. In fact, one may represent the map of gas bubbles, disregarding their shape and general behavior, on a plot of the Reynolds number and the bubble Eotvos number (Figure 3.7), where the Eotvos number is defined as

$$N_{Eo} = gd_e(\rho_L - \rho_G)/\sigma$$

A final example of systems where physical measurement using room-temperature liquids have been helpful in elucidating the behavior of a molten metals system is the field of gas jets impinging onto liquid surfaces. Bradshaw and Wakelin carried out extensive experiments studying the depth of the cavity generated by a gas jet impinging onto a liquid surface, as sketched in Figure 3.8, where H_0 is the distance from the jet orifice to the liquid level and H_c is the depth of the cavity.[9] The experimental results obtained (using the air–water and the mercury–CO_2 system) are shown in Figure 3.9, where the quantity $P_j = \pi d_0^2 \rho_G U_0^2/4$ is the jet momentum. It has been found that this relationship holds remarkably well for the BOF (LD) steel-making system.

In many instances one cannot construct a complete physical model of the system to be studied. *Under such conditions one must use engineering judgment to define the key components of the system* (e.g., flow through packed beds, agitation of liquids, using gas bubbles, impinging gas jets, and rising gas bubbles) and then proceed to characterize these through suitable experiments. The experimental conditions to be used need not satisfy all the similarity criteria, but it is helpful if the range of operating parameters, or at least those of the key dimensionless groups, does overlap that expected for the prototype.

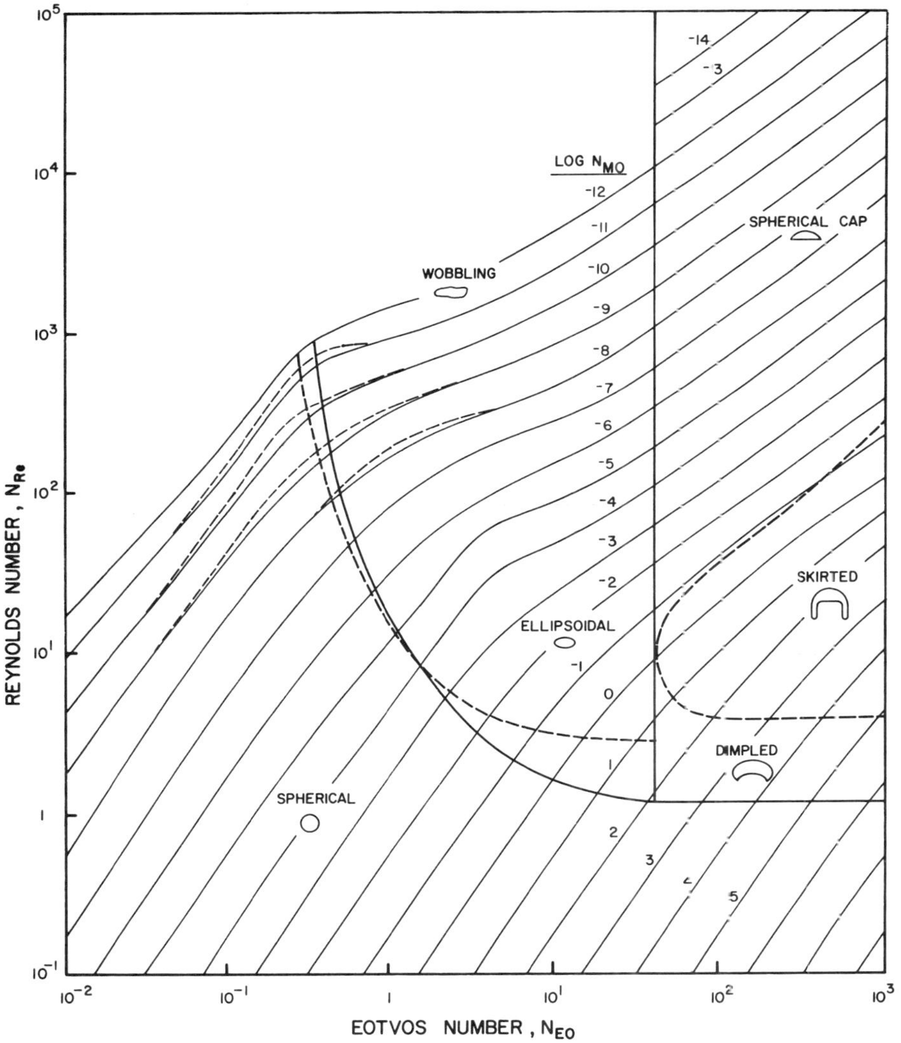

FIG. 3.7. Regions of bubble behavior on a plot of the Reynolds number against the Eotvos number, with the Morton number as a parameter after Grace *et al.*[11] Original figure kindly supplied by Prof. Grace of McGill University.

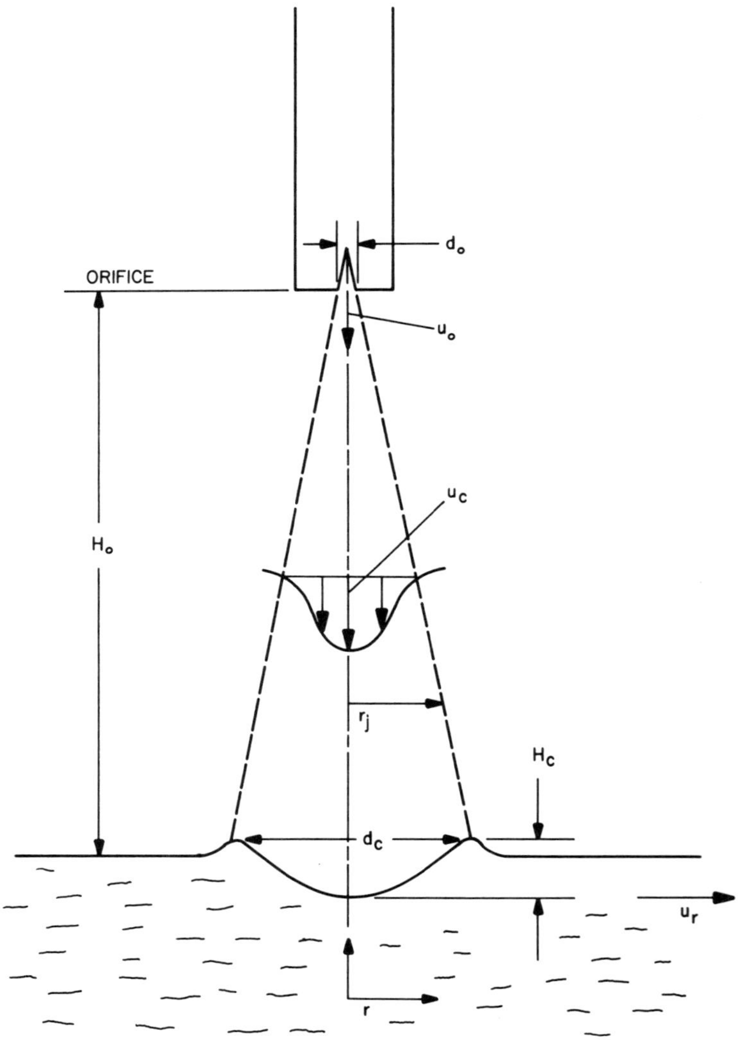

FIG. 3.8. The model of an impinging gas jet, after Bradshaw and Wakelin.[9]

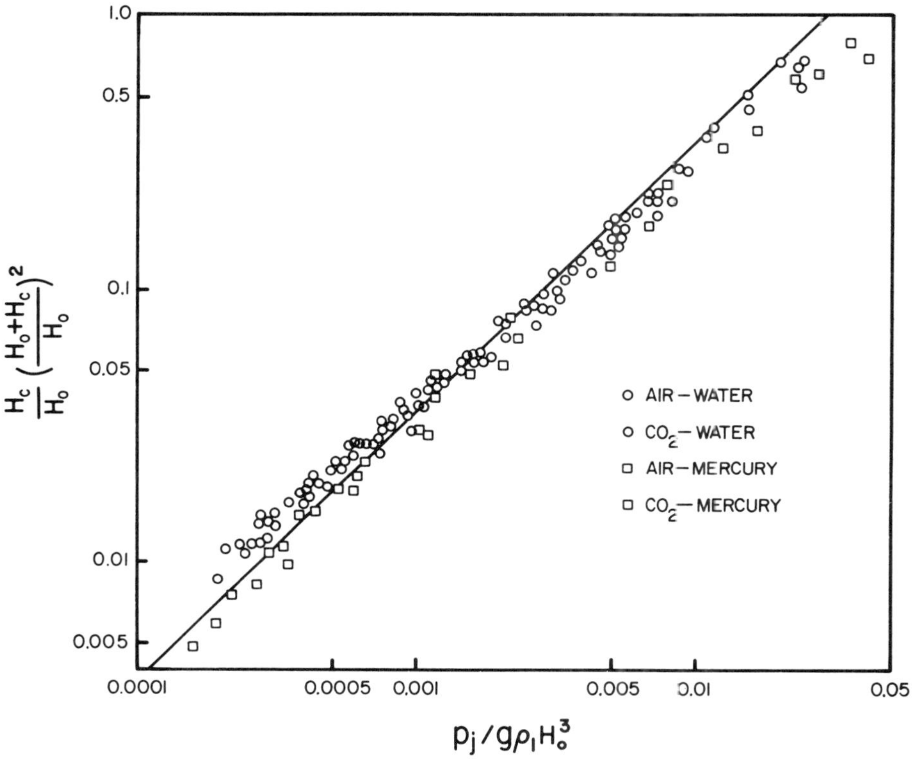

FIG. 3.9. The effect of jet momentum and nozzle height on the depth of the cavity produced when a gas jet impinges onto a liquid surface, after Bradshaw and Wakelin.[9]

3.4. PRELIMINARY AD HOC MEASUREMENTS

The development of semirigorous models or the detailed study of key physical phenomena might require quite an extensive commitment of time and resources. In some cases it may be helpful to conduct a preliminary investigation by performing some simple ad hoc measurements. Such measurements may be particularly helpful as a preliminary to mathematical model development when the investigator does not have a good physical feel for the system and the governing equations would have to be set up arbitrarily.

It is often tempting to construct a simple physical model of a system just to develop a qualitative understanding of how the key process parameters may be related. However, unless care is taken in the design of the experiments, the information thus obtained may be of little value.

In the design of ad hoc experiments, we have to start with a clear-cut definition of the objectives. For example, we wish to obtain a qualitative feel for the factors that govern the erosion patterns in AOD systems. In the definition of these objectives, it may be helpful to go through the exercise of formulating a mathematical model, thus identifying the missing relationships between the important process variables.

Then one may design an apparatus (taking advantage of all possible shortcuts, such as the availability of equipment), bearing in mind the similarity criteria. It is helpful if a geometric similarity is attained, or at least approached, and one must also be aware of how close one is to meeting the criteria for dynamic similarity.

There are no clear guidelines for planning preliminary ad hoc measurements, except for the fact that they are less rigorous than the experimental programs discussed in Sections 3.2 and 3.3. Nonetheless, careful planning, the clear definition of objectives, and a good level of awareness of the limitations are essential components.

3.5. CONCLUDING REMARKS

This chapter has introduced the concepts of physical modeling and outlined the rules for the development of rigorous physical models through the use of similarity criteria. It has also indicated, through examples, how these rules may be applied to the solution of practical problems.

While certain clear-cut rules govern physical modeling, this field is not an exact science. In many instances one has to rely on experience in the construction and development of physical models for novel processes.

However, numerous physical situations of general interest in metals processing have been well studied. In such cases, the construction of a physical model to represent a particular situation should be quite straight-forward. Examples of such situations include

- the behavior of gas bubbles in melts,
- gas bubble–stirred systems (see Chapter 6),
- pouring systems for tundishes and molds of continuous casting units,
- natural convection in ladles, and
- mechanically agitated systems.

For these situations, the appropriate dimensionless criteria are well defined, and the design of a model and the interpretation of the results should be an essentially routine task.

The physical modeling of novel systems poses a much more difficult problem, especially if dispersed phases are involved (e.g., sprays, swarms of bubbles, and slag–metal emulsions). For most complex systems, it is advisable to try to isolate the key phenomena, study them, and then synthesize an understanding of the system, rather than to seek to develop a separate model.

Physical modeling is just one tool that may be used in the characterization of metals processing operations. Ideally, the use of physical models, or of measurements to be carried out on model or relevant systems, should be part of an overall process development strategy. The extent to which physical modeling techniques must be employed depends on the nature of the system, the background information available on the system, and the mathematical skills of the investigator.

REFERENCES

1. *Symposium on Mathematical Process Models in Iron and Steelmaking*, Metals Society, London, 1973.
2. *Scaninject III: Proceedings of the International Conference on Injection Metallurgy*, MEFOS, Lulea, Sweden, 1983.
3. *Scaninject IV. Proceedings of the International Conference on Injection Metallurgy*, MEFOS, Lulea, Sweden, 1986.
4. J. Szekely, *Fluid Flow Phenomena in Metals Processing*, Academic Press, New York, 1979.
5. J. Szekely, J. W. Evans, and H. Y. Sohn, *Gas–Solid Reactions*, Academic Press, New York, 1976.
6. J. Szekely and N. J. Themelis, *Rate Phenomena in Process Metallurgy*, Wiley, New York, 1971, Chap. 16.
7. *Pilot Plants in the Iron and Steel Industry*, Special Report No. 96, Iron and Steel Institute, London, 1966. (See especially the paper by A. V. Bradshaw.)
8. R. E. Johnsone and M. W. Thring, *Pilot Plants, Models and Scale-Up Methods in Chemical Engineering*, McGraw-Hill, New York, 1957.
9. A. V. Bradshaw and D. Wakelin, in *Heat and Mass Transfer in Process Metallurgy*, A. W. D. Hills, ed., The Institution of Mining and Metallurgy, London, 1967.
10. M. Choudhary, J. Szekely, B. I. Medovar, and Yu Emelyanenko, *Met. Trans.* **13B**, 35 (1982).
11. J. R. Grace, T. Wairegi, and T. H. Nguyen, *Trans. Intn. Chem. Engrs.* **51**, 116 (1973).

4 REVIEW OF MEASUREMENT TECHNIQUES

This chapter summarizes the techniques available for measuring the most commonly encountered variables in mathematical and physical modeling. We have chosen to restrict our attention to velocity measurements, temperature measurements, and measurements of reaction rate. The reader will note that an important variable (concentration, or chemical composition) is omitted from this list. Numerous works in analytical chemistry deal with this field, and we believe that this province, with its rapidly developing instrumentation, is best left to others. The three quantities listed above are the variables of most interest to the process modeler.

We are concerned both with measurements on actual plants (where high temperatures, hostile conditions, vibration, etc. may make measurements difficult) and with measurements on physical models. Models are usually designed with the purpose of facilitating measurements that will test mathematical models, or at least provide information on the behavior of a process. Consequently, a physical model usually represents a compromise among the following needs:

To satisfy similarity criteria (described in Chapter 3),

To provide for ease of measurement,

To incorporate all the phenomena thought to be important in the real plant, and

To maintain costs of construction and operation of the model at a reasonable level.

Velocity measurements probably represent the greatest challenge to the modeler because of the multidimensional, unsteady nature of many flows encountered in metallurgy. Accordingly, this chapter devotes much attention to velocity measurement and attempts to clearly distinguish between those techniques applicable to actual plant measurements and those suitable only for physical models. No such distinction is appropriate for temperature measurement; indeed, there may be instances in which temperature measurements are simpler to make on the process than on a physical model (e.g., when the scale of the model results in significant heat losses by conduction along a thermocouple and a consequent temperature measurement error). Temperature measurements on models and on plants are therefore treated as one topic. Measurement of reaction rates is treated in terms of gas–solid and liquid–solid reactions, a major intention being to show the interaction of mass transport and reaction rates. Such interactions also are possible in gas–liquid and liquid–liquid reactions, and for such reactions the reader is referred to the literature.[1]

4.1. TECHNIQUES FOR VELOCITY AND PRESSURE MEASUREMENTS IN MODEL FLUIDS

Much physical modeling of metallurgical operations has been carried out using model fluids. Examples are the modeling of metal pools in continuous-casting units using water that was carried out by Szekely and Yadoya,[2] the use of kerosene to represent liquid metal in the metal filtration experiments of Mutharason and Apelian,[3] and the use of mercury in the work of Tarapore and Evans on induction furnaces.[4] The use of such model fluids, rather than the real fluids of interest, is due to the much greater ease with which the fluids may be handled in a laboratory and the fact that in many instances measurement of such quantities as velocity within the real fluid is difficult or impossible.

This section is therefore devoted to measurements within model fluids. Since such fluids are frequently water, aqueous solutions, or common gases such as air, the techniques reviewed here are those familiar to most fluid mechanicians.

The *static tube* illustrated in Figure 4.1 is an obvious and well-known device

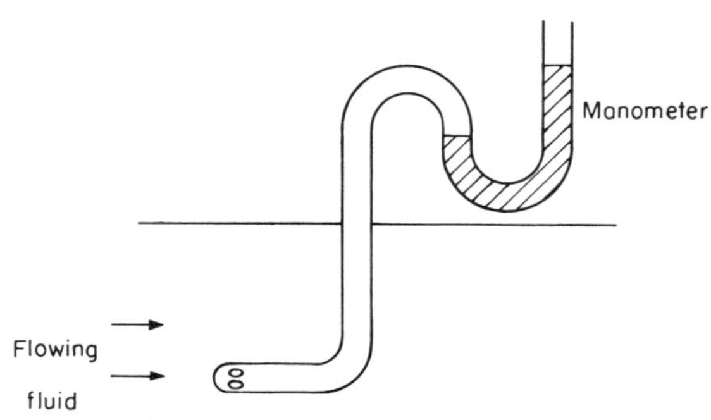

FIG. 4.1. The static tube for pressure measurement in a flowing fluid.

for measurement of fluid pressure. Usually the pressure measured by the device is not independent of the fluid velocity, and it is wise to check the insensitivity by a calibration experiment. This is particularly true if there is a danger that the tube may be yawed to the direction of the flow. Pressure within the fluid can be read from the simple U-tube manometer depicted in Figure 4.1 or from an inclined manometer if greater sensitivity is required. Other even more precise pressure measurement devices, such as capacitance gauges, are commercially available if greater precision is required.

The *Pitot-static tube* depicted in Figure 4.2 is a device for simultaneous measurement of local velocity and pressure. Application of Bernoulli's equation yields an equation relating the velocity to the reading on the manometer:

$$V = C\sqrt{2gh(\rho_0/\rho - 1)}$$

where C is a correction coefficient (close to unity) best determined by calibration. The Pitot-static tube is insensitive to yaw over an angle of a few degrees; this has both beneficial and detrimental consequences. If the local flow direction is known, then the Pitot-static tube can be used to measure the magnitude of the velocity (and the local pressure) without the necessity of taking elaborate precaustions to align the tube correctly. However, if the local velocity direction is *not* known, then the Pitot-static tube cannot be used to ascertain the direction. The signal from a Pitot tube is very sensitive to yaw angle at about 45°; consequently, the device depicted in Figure 4.3 can be used to measure local flow direction before measurement of velocity with a Pitot-static tube. The pressure reading on the manometer is zero when the velocity direction is midway between the two tube directions. The principle can be readily extended to three-dimensional flows.

The Pitot-static tube is a relatively inexpensive device for velocity measurement. Perhaps the next most inexpensive techniques are the photographic

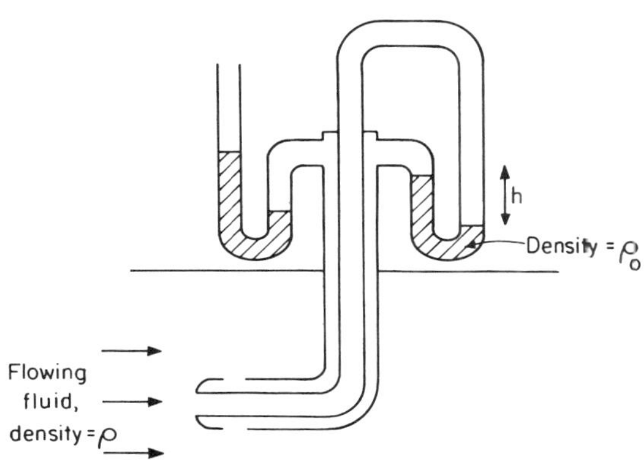

FIG. 4.2. The Pitot-static tube for measurement of velocity.

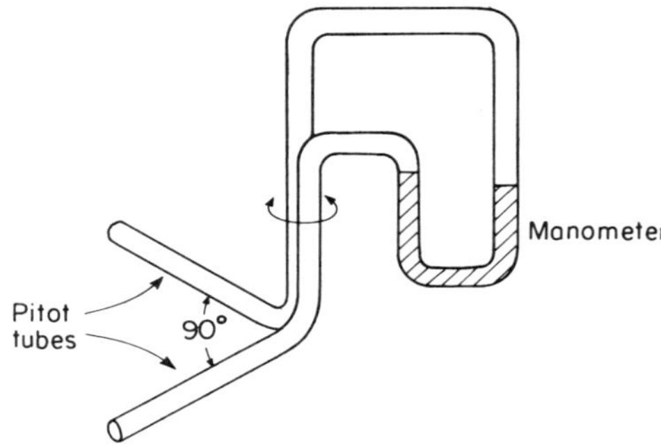

FIG. 4.3. Device using Pitot tubes to determine flow direction.

methods of velocity determination. Such methods can also provide a rapid graphic representation of various aspects of the flow field; for example, they can reveal the presence of regions of recirculating flow.

Photographic methods have been treated at length.[5] Model fluids are usually transparent, and the photographic methods consist of introducing some foreign material into the flow and recording the movement of the material. Dye, smoke, and vapor introduced into the fluid can give a visualization of the flow direction in the bulk of the fluid. Tufts attached to surfaces can show flow directions at these surfaces. However, none of the above is suitable for velocity measurement.

Particles or bubbles introduced into the fluid provide a ready means of velocity measurement. For gas flows a device for generating helium-filled soap bubbles is commercially available. The device is capable of adjustment so that the bubbles are of neutral buoyancy in the gas of interest. The simplest such technique is to photograph the particles using an exposure that is sufficiently long that the resulting negative or print shows the particles as short streaks. Measurement of the length of the streaks enables the calculation of the velocity at that point, given the exposure time. The following difficulties are associated with this technique: (1) optical problems such as refraction and parallax, (2) exposure errors, and (3) mismatch between particle and fluid velocity.

The first of these difficulties is easy to overcome by minimizing the curved or inclined surfaces through which the light beams pass en route to the camera and by illuminating only a small region of the fluid (e.g., a plane normal to the line of sight of the camera), perhaps including a scale directly in this region. The second difficulty arises because the mechanical shutters on most commercially available cameras are imprecise. Frequent calibration of the shutter may be required, but a simpler approach is to use an electronically controlled stroboscope for illumination, whereupon the particle trajectories appear as a line of dots. If the shutter is left open for long enough under stroboscopic lighting, then the dot-to-dot distances give the velocity profile along the trajectory of the particles. Figure 4.4 is a print made in this way of a glass–tantalum float moving across the top of an inductively stirred mercury pool.

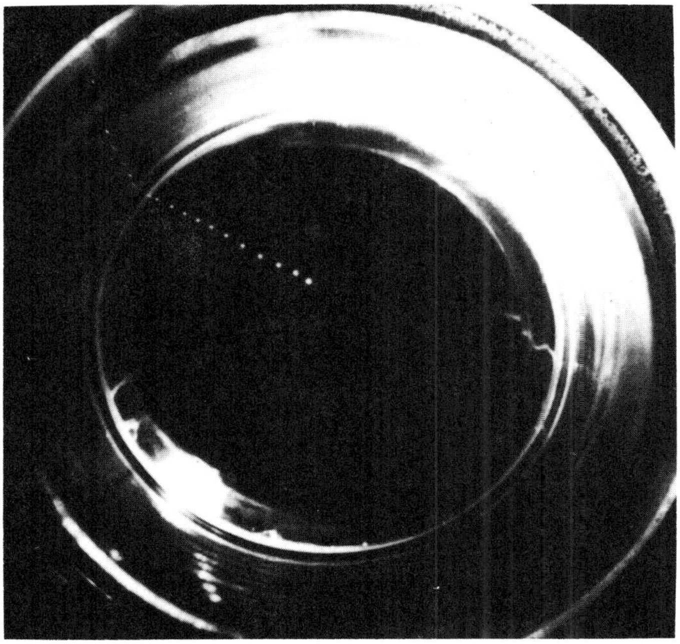

FIG. 4.4. Time exposure photograph, under stroboscopic lighting, of a float on an inductively stirred mercury pool.

The third difficulty is a serious one. Consider a particle in a turbulently flowing liquid. If the particle is large it will be incapable of "keeping up with" the mean (time-averaged) velocity of the fluid. For example, if the fluid streamlines curve, there will be a tendency for the particle to pursue a straight line instead (if the particle is more dense than the fluid). Furthermore, such large particles have significant terminal velocities at which they sink or float within the liquid, introducing further error. If the particle is smaller, it may be "in step" with the mean velocity, but there is a danger that if the particle is too small it may become responsive to the turbulent fluctuations of velocity. In the latter case the photographic technique will yield an *apparent* lack of reproducibility, and it will be necessary to average the results of a large number of velocity measurements at a point in order to get the time averaged velocity. In theory even smaller particles should follow the instantaneous velocity, and photographic techniques might then yield information on turbulence. Hjelmfelt and Mockros[6] show, by a numerical solution of the equation governing particle motion in a fluid (the Bassett–Boussinesq–Oseen equation), that even quite small particles cannot keep up with the turbulent fluctuations. They identified a frequency for turbulent fluctuations above which the particle would deviate significantly form the flow. This frequency appears in Table 4.1 for particles, droplets, and bubbles such as might be used in the photographic technique. Since the spectrum of

TABLE 4.1. Frequencies Limiting the Accuracy of Various Particle–Fluid Combinations for Velocity Measurement in Turbulent Flow

Particle and Fluid	Diameter (μm)	Limiting Frequency (Hz)
Sand and water	500	0.16
Water droplets and air	7	812
Hydrogen bubbles and water	200	20

frequencies in turbulent fluid flows extends up to about 10,000 Hz, the photographic technique is not suitable for turbulence measurements when used on the above particle–fluid combinations. The very small particles used in laser velocimetry (see Section 4.1.3), which do follow the turbulent fluctuations, are too small for normal photography.

The foregoing discussion has concerned the use of a still camera, but similar measurements can be made with a movie or video camera. In this case, the frame-to-frame movement of the particles is determined by means of a stop-action projector or a frame analyzer. Calibration of the camera by, for example, taking footage of a stopwatch is usually desirable.

4.1.1. Hot-Wire and Hot-Film Anemometry

A simplified electrical circuit for a hot-wire (or hot-film) anemometer appears in Figure 4.5. The hot wire is a narrow-gauge wire whose resistance is temperature sensitive. The wire is immersed in the flowing fluid, and heat generated in the wire by the electric current is dissipated into the fluid. Consider an increase in fluid velocity. The heat generated in the wire remains approximately constant and, at steady state, equals that transferred to the fluid. The increase in velocity increases the heat transfer coefficient from wire to fluid, lowering the temperature. The change in wire temperature yields a change in resistance, which in turn leads to a change in the measured voltage across the bridge. With

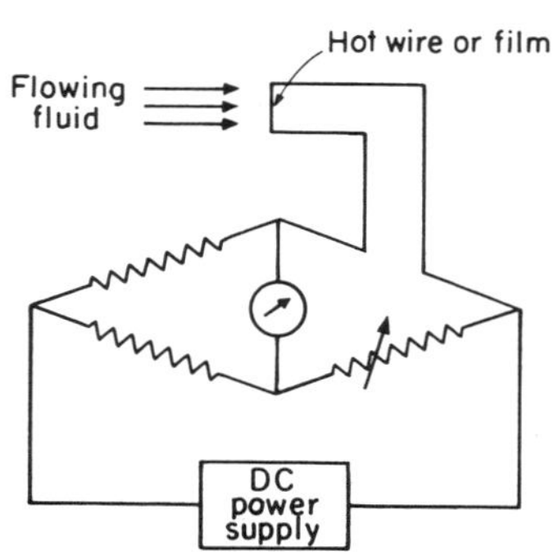

FIG. 4.5. Simplified electrical circuit of a hot-wire anemometer.

sufficiently fine wire and appropriate circuitry for processing the voltage signal from the bridge, the hot-film anemometer can be used to determine turbulence levels as well as to measure mean stream velocities. The hot-wire anemometer is too fragile for use in liquids, and in such cases the hot-film anemometer is used. The principle of operation is the same, but now the wire is replaced by an electrically conducting thin film on a suitable substrate The performance of either instrument can be impaired by dust or bubbles in the fluid. These devices respond to fluctuations in fluid *temperature* as well as velocity, but compensated probes are available wherein, for example, a second resistance wire, alongside the hot wire but not electrically heated, provides input to the electronic circuitry concerning the local temperature. It is necessary to calibrate these anemometers in the fluid to be used. Saunders and Lawrence[7] have described a suitable calibration machine for hot-film anemometers over the range 10 mm/s to 3 m/s. The calibration of hot-wire anemometers can be achieved by using a pitot-static tube or by using a total flowmeter (e.g., a rotameter). In the latter case the probe is inserted in the center of a cylindrical pipe down which the gas is passing in laminar flow. The velocity at the probe is then twice the average velocity obtained from the flow meter and the pipe cross section.[8] The results of calibration experiments are frequently correlated by the King equation:

$$E^2 = a + bU^n$$

where E is the voltage across the probe,
U is the velocity, and
a, b, n are constants.

Davies and Patrick[9] have suggested an alternative equation that they claim yields a superior fit.

The sensitivity of these anemometers to temperature fluctuations, dirt, and bubbles has already been mentioned. Additional difficulties are their rapid deterioration in corrosive fluids and directional problems. A hot wire is mainly responsive to the velocity component perpendicular to the direction of the wire but unfortunately is not insensitive to flow parallel to the wire. Consequently there is a directional ambiguity about the results provided by the instrument. This can be partially resolved by changing the probe orientation until the maximum velocity is measured; the flow will then be perpendicular to the wire. Unfortunately this still leaves the flow direction ill defined and either a cross-type probe, equipped with two perpendicular wires, or auxilliary measurements (e.g., using the device in Figure 4.3) must be employed to determine the flow direction. Further directional ambiguities arise if the probe is used to measure turbulence intensities. One technique for unambiguously determining the direction of the fluid has been described by Taberlet.[10] Downstream of a hot-film probe is a wake of heated fluid, and Taberlet used a second probe to locate the position of this wake, thereby defining the direction of the fluid (in this case an inductively stirred pool of mercury).

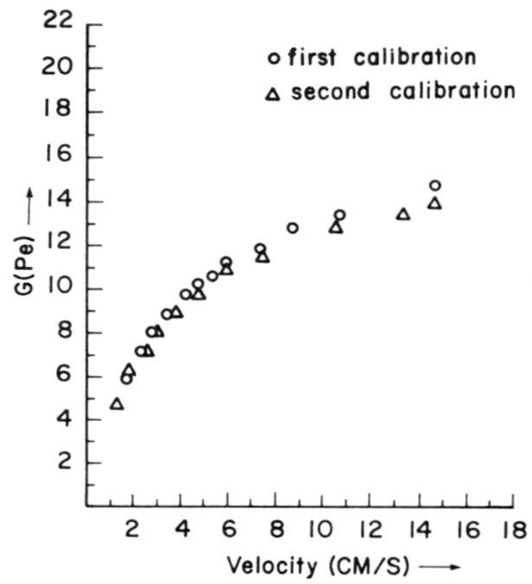

FIG. 4.6. Calibration curves for a hot-film velocimeter in Wood's metal.

Hot-film probes have been widely used for velocity measurements in mercury.[11–13] In such applications it is necessary that the electrically heated film be insulated from the mercury (usually by a quartz film). Recently Murthy and Szekely[14] have applied the hot-film probe to velocity measurements in Wood's metal. Both mercury and Wood's metal are oxidized by air, and despite protective blankets of inert gas it is common to encounter drift of the signal due to fouling of the probe with oxide. This difficulty can be mitigated by correcting the probe voltage by the change of the probe signal in stationary liquid. This is achieved (following the method of Sabjen[11]) by use of the function

$$G = \left(\frac{1}{E_0^2} - \frac{1}{E_T^2}\right)\phi$$

where E_0 is the probe voltage in stationary liquid (measured as soon as possible before use of the probe in moving liquid),

E_T is the voltage in moving liquid, and

ϕ is an arbitrary scaling constant.

Figure 4.6 shows a plot of G versus velocity obtained by Murthy and Szekely for two calibration experiments in Wood's metal separated by 2 days. It is anticipated that such probes (or the electromagnetic probe of Vives described in Section 4.1.2) will find wide use in physical models of metals processing operations.

4.1.2. Electromagnetic Velocity Measurements

The principle of electromagnetic velocity measurement has been used for some time in industry for measuring total flows. However, its application to local velocity measurements has been limited. Consider Figure 4.7, where a fluid is

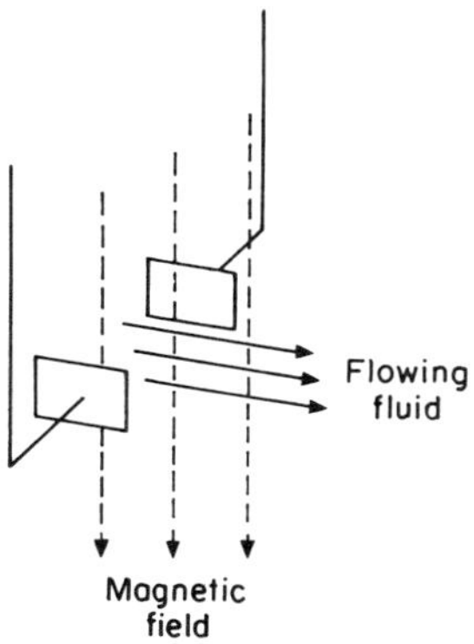

FIG. 4.7. Principle of a magnetohydrodynamic velocity probe.

flowing between electrodes in a direction perpendicular to an applied magnetic field.

According to Ohm's law for a moving conductor (discussed in Chapter 2),

$$\mathbf{J} = \sigma(\mathbf{E} + \mathbf{V} \times \mathbf{B})$$

where $\mathbf{J}$ is the current density between electrodes,
$\mathbf{E} = \nabla\phi$ is the electric field,
ϕ is the electric potential,
$\mathbf{V}$ is the velocity vector,
$\mathbf{B}$ is the magnetic induction vector, and
σ is the electrical conductivity.

If no current flows and $\nabla\phi$ can be approximated by $\Delta\phi/\Delta x$, where $\Delta\varphi$ is the potential difference between electrodes and Δx is the electrode separation, then

$$\Delta\phi = \Delta x\, VB$$

and measurement of potential across the electrodes yields the instantaneous local velocity between them in the direction perpendicular to the field. Although the electrical conductivity of the fluid does not appear in this equation, in practice it is necessary for the fluid to have some conductivity; in this way measurement of the potential between the electrodes with a voltmeter of normal resistance yields the desired open-circuit voltage. The principal difficulty with this technique is the small voltages generated (of the order of a few tens of microvolts). Janalik[15] has used the technique for measurement of turbulence intensities in pipe flows. Figure 4.8, taken from this work, shows reasonable agreement between magnetohydrodynamic measurements (MHM), hot-film anemometry (CTA), and the correlation of Martin and Johanson.[16]

An interesting recent development is the incorporation of a permanent

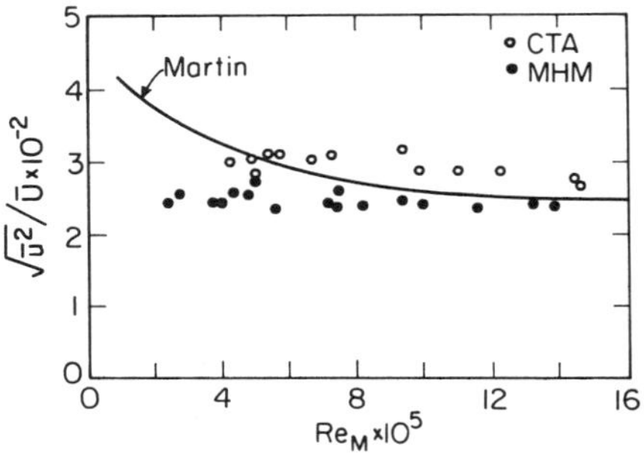

FIG. 4.8. Turbulence intensities in pipe flows measured using the magnetohydrodynamic and hot-film techniques, after Janalik.[15] Reprinted with permission.

magnet into a probe of this electromagnetic type.[17] Two "electrodes" (the exposed tips of otherwise insulated wires) are placed antipodally in the midplane of a cylindrical permanent magnet a few millimeters in diameter. When this probe is placed in a flowing conducting fluid, a voltage develops between the electrodes. This voltage is proportional to the velocity of the fluid and independent of the physical properties of the fluid. By incorporating two more electrodes in the midplane (on the perpendicular of the line joining the first two electrodes), two velocity components can be measured simultaneously. Furthermore, the probe is responsive to the sense of the flow, that is, the voltage difference changes sign when the flow reverses direction.

The Figure 4.9 below shows the results of calibration experiments by Ricou and Vives.[17] The linear response and lack of dependence on the metal should be noted. The decline in the response with increasing temperature is a consequence of the decline of the strength of the permanent magnet as the Curie temperature is approached. These workers found that the response at a given temperature was reproducible (i.e., was the same on returning to that temperature from other

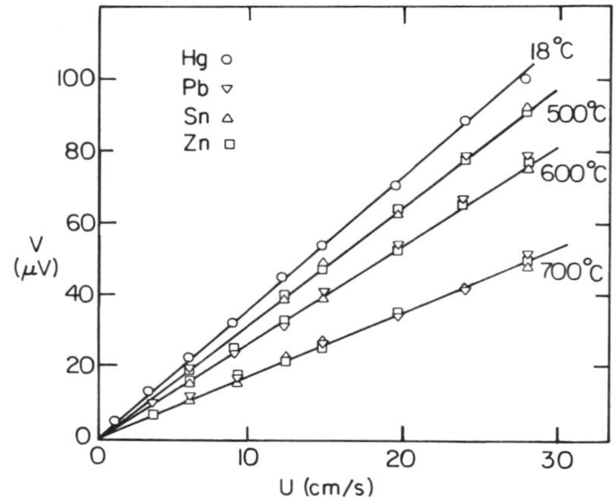

FIG. 4.9. Calibration curves at different temperatures for the magnetohydrodynamic probe of Ricou and Vives.[16] Reprinted with permission from *International Journal of Heat and Mass Transfer, vol. 25*, R. Ricou and C. Vives, "Local Velocity and Mass Transfer Measurements in Molten Metals Using an Incorporated Magnet Probe", 1982, Pergamon Press Ltd.

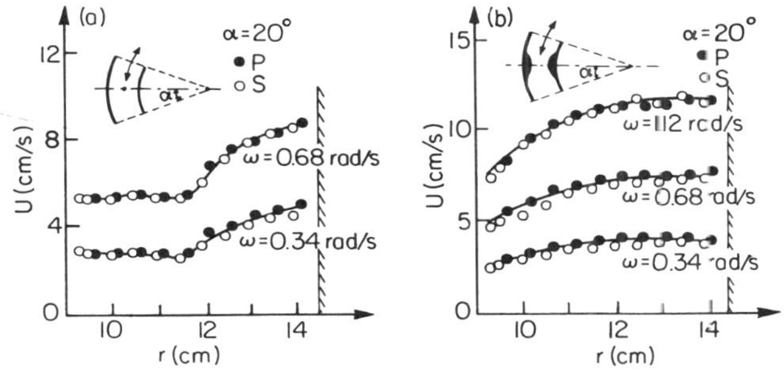

FIG. 4.10. Comparison of velocity measurements made by the magnetohydrodynamic probe of Ricou and Vives (S)[16] and by Pitot tube (P) (*a*) downstream from a small cylinder and (*b*) downstream from a widening. Reprinted with permission from *International Journal of Heat and Mass Transfer*, *vol. 25*, R. Ricou and C. Vives, "Local Velocity and Mass Transfer Measurements in Molten Metals Using an Incorporated Magnet Probe", 1982, Pergamon Press Ltd.

temperatures). Furthermore, there was virtually no drift with time on holding the probe at an elevated temperature.

Figure 4.10 shows a comparison of velocity measurements made by the incorporated magnet probe and those made by a Pitot tube in mercury. In unsteady flows the probe showed an instantanous response over the frequency range studied (up to 50 Hz), and the probe should therefore be suitable for some turbulence studies in metals when the length scale of the turbulence is greater than the probe diameter. The main disadvantage of the probe is its weak output signal (of the order of microvolts), requiring expensive electronic equipment for measurement and precautions to avoid noise, thermal emfs, and so on. Furthermore, difficulties may be encountered when an external electric or magnetic field is applied to the melt (e.g., when measuring electromagnetically driven flows). Lee, Evans, and Vives[18] have recently described the use of the probe to measure velocities in Wood's metal.

4.1.3. Laser–Doppler Velocimetry

The laser–Doppler velocimeter (LDV) is the most sophisticated tool for measurement of velocity components and turbulence levels in transparent fluids. In this chapter, there is space for only scant coverage of this instrument, and the interested reader is referred to a book on this topic.[19]

The instrument is illustrated in Figure 4.11. A laser beam is split into two

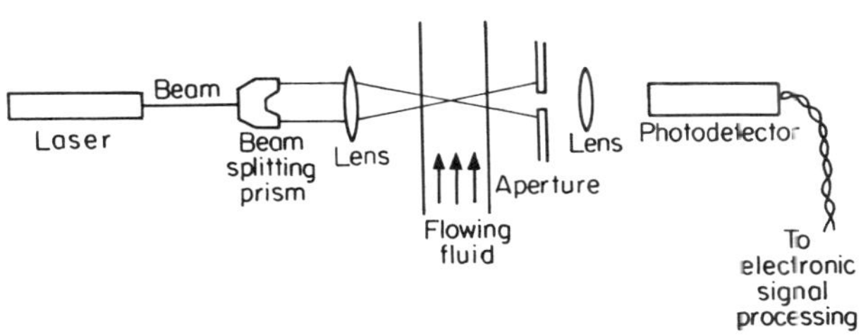

FIG. 4.11. Simplified drawing of a laser–Doppler velocimeter.

beams by a prism, and a lens causes the two parts to cross at the point in the fluid where velocity is to be measured. A simplified statement of how the instrument works is as follows. The intersecting beams set up an interference pattern at the intersection point. This pattern, as seen along the direction of the optical axis, is a series of bright and dark bands, which lie perpendicular to the plane in which the intersecting beams lie. Imagine a small particle borne along by the fluid. As the particle passes through the interference pattern, it will be illuminated by light of varying intensity. Some of this light is scattered by the particle into the receiving optics, which consist of an aperture (to screen out the direct beams), a photodetector (to measure the scattered light intensity), and a lens (to focus the scattered light onto the photodetector).

The output of the photodetector is sketched in Figure 4.12 for one particle passing through the interference fringes. The frequency of the signal from the photodetector is proportional to the velocity of the particle and inversely proportional to the fringe spacing d, which is given by

$$d = \frac{\lambda}{2 \sin \theta}$$

where θ is the half angle between the beams. For typical lasers and beam angles, d is of the order of a few micrometers. For a velocity of 10 cm/sec this yields a frequency of the order of 100 kHz. This signal is fed into electronic processing equipment that is essentially a frequency measurement device. The velocity measurement is completed in a time period of the order of 10^{-5} s and is therefore virtually an instantaneous measurement. Consequently, with appropriate electronic equipment, the LDV can yield information on turbulence quantities as well as the time-averaged velocity. This description of how the LDV operates is a simplistic one. A better description is that the device measures the beat frequency between the two scattered light signals. The scattered light signals are the incident laser beams that have been Doppler shifted in frequency by the motion of the particle. The frequency shifts are different for the two beams, since the velocity vector of the particle makes different angles with each of the two incident beams.

It is clear that, as described, the instrument is sensitive only to the velocity component lying in the plane of the two beams. The other two components yield

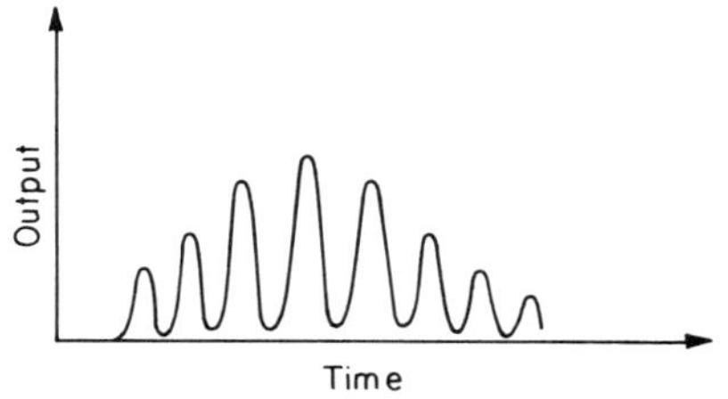

FIG. 4.12. Output from the photodetector of a laser–Doppler velocimeter.

no signals. This is an advantage in that the direction of the velocity being measured is defined. Unfortunately, there is still an ambiguity in that a particle moving up through the fringes with a certain speed generates the same signal as one moving down with that speed. This ambiguity can be overcome by installing a frequency shifter in one of the beams. The effect of frequency-shifting one of the beams is to make the fringe pattern sweep downward at a known speed that makes stationary particles give a signal of nonzero frequency, upward moving particles a frequency that is higher, and downward moving particles a frequency that is lower.

In the apparatus depicted in Figure 4.11, it is a relatively simple matter to measure the two velocity components perpendicular to the optical axis. This entails merely rotation of the prism. Of course, this does not permit simultaneous measurement of these components, such as might be important in turbulent flows. Such simultaneous measurements are possible in more complicated optical arrangements using four laser beams in two perpendicular planes and all intersecting at the same point. Each pair of laser beams is differentiated by its polarization or its color (using two lasers), with two photodetectors being used. More complicated systems make possible the measurement of all three velocity components simultaneously.

The LDV shown in Figure 4.11 is operating in the dual beam, forward scatter mode. Operation in this mode is common. Backscatter operation involves mounting the photodetector on the same side of the flow as the laser. The reference beam mode of operation is one employing incident beams of differing intensity, the weaker of which enters the photomultiplier directly together with scattered light from the stronger beam; this mode finds application when the transparency of the fluid is limited.

A discussion was provided earlier of the need to employ small particles when using the photographic technique for velocity measurement. This need also arises with particles used in laser velocimetry. Table 4.2 lists some particles used in various fluids. Appearing in this table are entries for the maximum particle size that can keep up with a sinusoidal velocity fluctuation of 1 kHz and 10 kHz.

TABLE 4.2. Particles Used in Laser Velocimetry

		Diameter (μm)	
Particle	Fluid	1 kHz	10 kHz
Silicone oil	Air	2.6	0.8
TiO_2	Air	1.3	0.4
MgO	Methane–air flame, 1800 K	2.6	0.8
TiO_2	Oxygen plasma, 2800 K	3.2	0.8
PVC	Water	16	5

SOURCE: Reference 19.

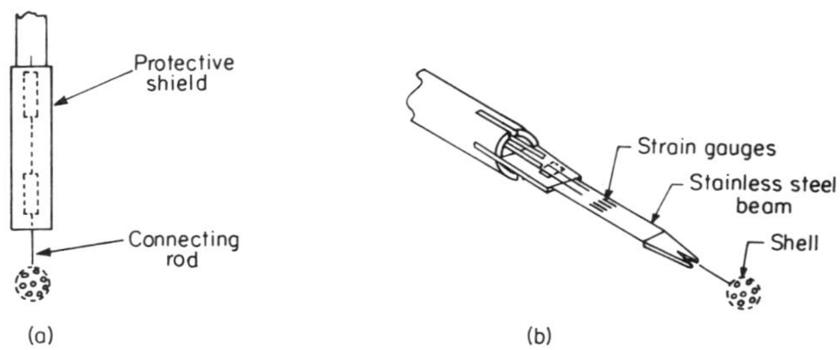

FIG. 4.13. Velocity probe of Moore and Hunt.[21] Reprinted with permission from Keter Publishing House Jerusalem Ltd.

Particles of this size are available and scatter sufficient light for an acceptable signal from the photomultiplier. Particle generators are commercially available.

Frequently there are enough small particles in tap water to make seeding unnecessary when studying water flows. Care must be taken to avoid overseeding the flow. If more than a few particles pass through the fringe pattern at once, then the light scattered from each interferes with that from the others and the photodetector generates an unacceptable signal.

It is appropriate to compare the LDV with other methods of velocity (and turbulence) measurement. According to Humphrey,[20] LDV offers "spatial resolution an order of magnitude better than that of a hot-wire anemometer, directional sensitivity and easy measurement of two- and three-dimensional flows. In addition, the instrument does not disturb the flow, measures at a point, and requires no calibration."

4.1.4. Velocity Measurements Using a Reaction Probe

Moore and Hunt[21] have recently developed a velocity probe based on measuring the force exerted on a submerged sphere by the flowing fluid. This is depicted in Figure 4.13. A spherical perforated tantalum shell is attached by means of a connecting rod to strain gauges capable of measuring the force component perpendicular to the probe axis. The response of the probe is sufficiently rapid that turbulence measurements can be made. Velocities measured by Moore and Hunt in an inductively stirred mercury pool and a comparison with theory are presented in Chapter 6. A more recent version of this probe[22] permits the simultaneous measurement of two velocity components.

4.2. VELOCITY MEASUREMENTS IN MOLTEN METALS, SLAGS, AND SALTS

Difficulties are encountered in applying the methods discussed in Sections 4.1.1–4.1.4 to molten metals, slags, and salts. The Pitot-static tube suffers from the difficulty of coupling it to the pressure sensor without freezing liquid in the lines or overheating the sensor. Opacity of the liquids rules out the photographic techniques and laser velocimetry (except for measurements at free surfaces), although Sadoway and Szekely[23] have used laser velocimetry to measure velocities in a molten salt. Hot-film anemometers are unable to withstand the high temperatures and corrosive nature of most liquids encountered in pyrometallurgical operations.

4.2.1. Surface Velocity Measurements

Tarapore et al.[24] measured surface velocities on a 12-ton steel melt in an induction furnace by measuring frame-to-frame movement of slag globules in a movie film of the melt surface. Szekely et al.[25] carried out similar measurements on a 30,000-lb melt using a video recorder.

In discussing photographic methods of studying pyrometallurgical operations, recent work using tuyerescopes should be mentioned. Bustos et al.[26] have used a movie camera attached to a tuyere of a copper converter or a zinc fuming furnace to study gas injection phenomena (particularly the formation of accretions at the tuyere) in nonferrous operations. Another outstanding example of this kind of work is that of Kase and coworkers,[27] who used a high-speed camera to make observations through the tuyere of a blast furnace. Coke particle size distributions, velocities, and raceway sizes were obtained in this study. Similar work was later carried out by Yamaguchi et al.[28] using an even faster camera coupled to an image analyzer.

4.2.2. Reaction Probes

Figure 4.14 illustrates a reaction probe used by Szekely et al.[29] to measure melt velocities in Wood's metal contained within an induction coil. The reaction probe consisted of a disk supported on a vertical rod. The rod was held in place by a spring and its vertical movement measured by a differential transformer. The principle of operation of the probe is akin to that of the Pitot tube in that the device responds to the stagnation pressure developed when fluid impacts the disk. By application to Bernoulli's equation, the fluid velocity just before impact with the disc can be related to the force on the disk by

$$V = \sqrt{2F/A}$$

where A is the area of the disk and F, the force, is related to the output of the differential transformer by prior calibration experiments using a top-loading balance. Corrections for buoyancy effects on the probe are necessary. Furthermore, this equation neglects viscous drag on the probe, and, for precise measurement, calibration experiments in flowing metal would be required. As shown in Figure 4.14, the probe can measure only vertical velocity components, but there should be no difficulty in extending the technique to measure velocities

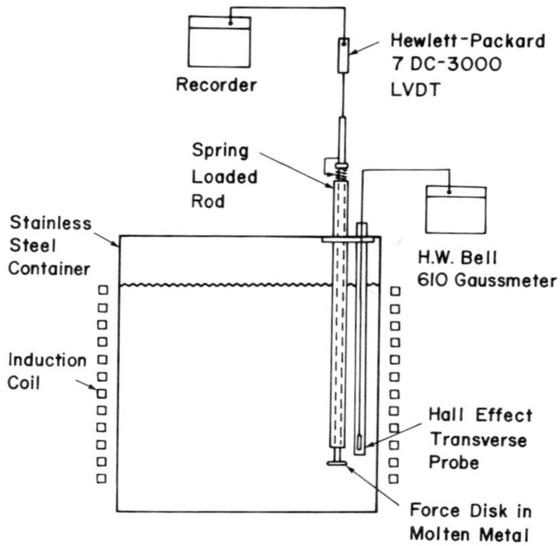

FIG. 4.14. Measurement of metal velocities using a reaction probe.[29]

in other directions, as Moore and Hunt[21] have done for mercury in the work described in Section 4.1.4. A similar reaction probe has been used recently by El-Kaddah et al.[30] to measure velocities in an inductively stirred steel melt.

4.2.3. Velocities from Dissolution Rate Measurements

Johnson[31] has described an interesting technique for the measurement of metal velocities in the aluminum pool within the Hall–Héroult cell. An iron rod is held in a vertical position with its lower end in the aluminum pool for a measured period of time (typically 10–20 min). The iron dissolves into the aluminum under mass transport control (i.e., at a rate that is dependent on the local velocity of the aluminum). Calibration experiments were carried out by towing rods through stationary aluminum held in a ladle; Figure 4.15 is the result of the calibration experiments. The technique yields a great deal of information. Figure 4.16 is a sketch of the appearance of such rods after immersion. Variation of the dissolution rate from top to bottom of the aluminum pool enables the determination of a velocity profile, not merely a velocity. Furthermore, dissolution of the rod occurs more rapidly from the upstream edge, enabling the velocity directions to be fixed. Velocity measurements made in this way are reported in Chapter 8.

In principle, velocities in the molten cryolite electrolyte could be measured in a similar way (e.g., by measuring the rate of dissolution of an alumina rod). In practice this is difficult because of the small difference between the molten salt temperature and its freezing point (10–20°C compared to about 300°C for the aluminum), which causes considerable freezing of cryolite onto the rod. A

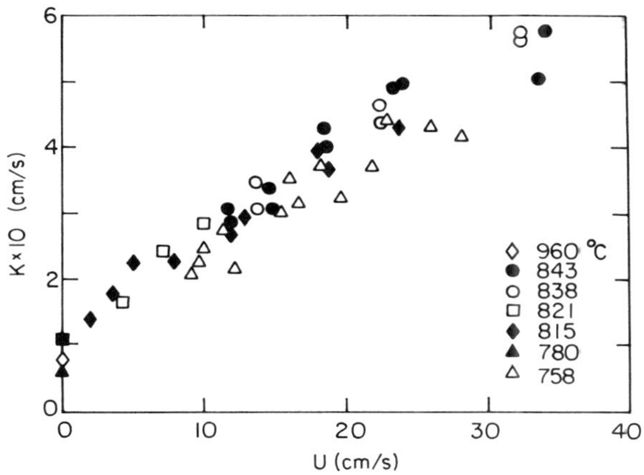

FIG. 4.15. Measurements of Johnson on the rate of dissolution of an iron rod in aluminum at various velocities.[31]

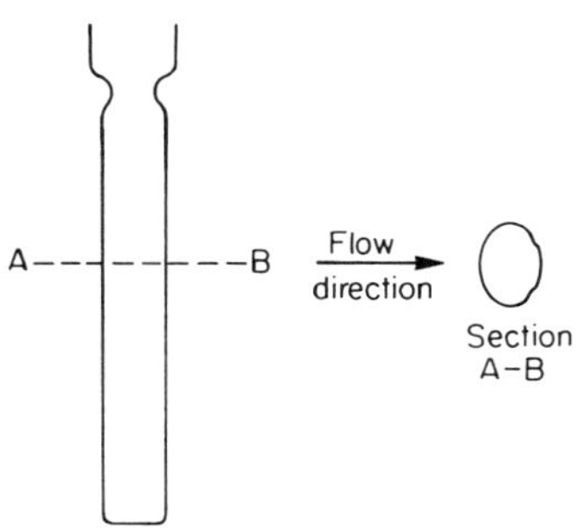

FIG. 4.16. Shape of iron rod after partial dissolution in flowing aluminum.

further difficulty is the sensitivity of the dissolution rate to the alumina content of the cryolite. Rather similar studies of fluid flow and mass transfer in a steel melt have been carried out by El-Kaddah et al.[30] using graphite rods.

4.2.4. Tracer Studies

A technique that has found use for velocity measurement in molten salts and metals is the tracer method. A small amount of tracer (usually a radiotracer) that is completely miscible with the melt is introduced at some point, and the melt is sampled periodically at a downstream location. The technique can yield information on the turbulent diffusivity of the melt as well as on the velocity between the point of tracer introduction and the sampling point.

Consider the simple flow system depicted in Figure 4.17, where at $t = 0$ the tracer is introduced into a melt at a point that is a distance d upstream from the sampling point. Let us adopt a radial coordinate system that moves with the

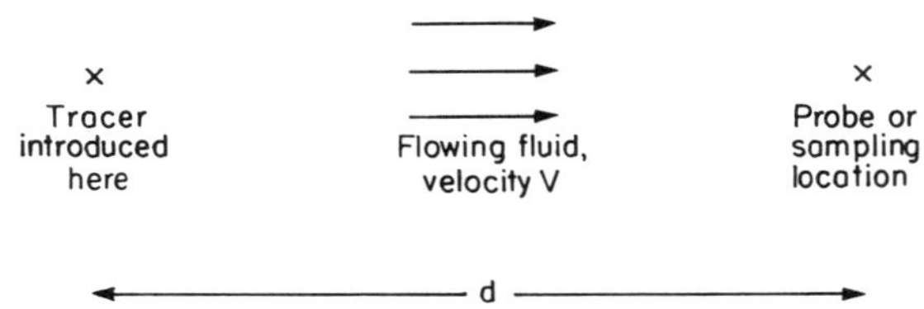

FIG. 4.17. Tracer technique for measurement of velocity and diffusivity.

fluid and has as its initial origin the point of introduction of the tracer. Assuming isotropy and that the decay of the tracer is negligible over the time period of the experiment, the concentration of the tracer is given by

$$\frac{\partial c}{\partial t} = \frac{D_t}{r^2}\frac{\partial}{\partial r}\left(r^2\frac{\partial c}{\partial r}\right)$$

where D_t is the diffusivity in laminar flow or the effective diffusivity in turbulent flow.

The boundary and initial conditions are

$$\begin{aligned} &\text{as} \quad r \to \infty, \quad && C \to 0 \\ &\text{at} \quad r = 0, && C \text{ is finite for } t > 0, \qquad \text{and} \\ &\text{at} \quad t = 0, && C = \delta(0) \end{aligned}$$

The first boundary condition assumes that the volume of melt over which the tracer is dispersed is small compared to the total melt volume, while the initial condition supposes the volume into which the tracer is initially introduced is negligibly small.

These equations have the solution

$$C = \frac{A}{2\sqrt{\pi D_t t^3}} \exp\left(\frac{-r^2}{4D_t t}\right)$$

where A is a constant.

In the moving coordinates system, the location of the sampling device or probe is given by

$$r_{\text{probe}} = d - Vt$$

where it is assumed that a constant melt velocity exists between the point of tracer introduction and the probe. Consequently, the concentration measured at the probe point is

$$C = \frac{A}{2\sqrt{\pi D_t t^3}} \exp\left(-\frac{(d - Vt)^2}{4D_t t}\right)$$

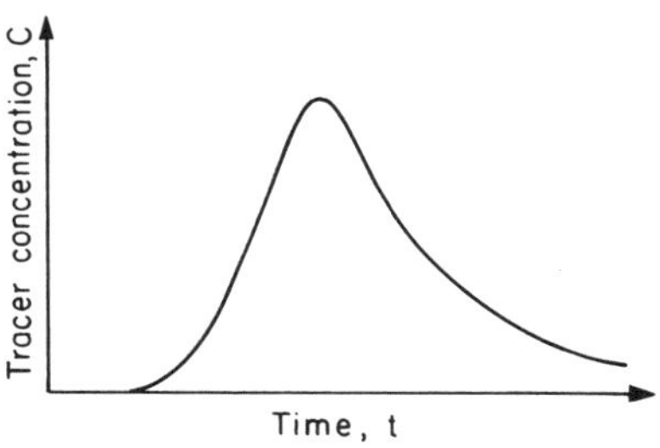

FIG. 4.18. Tracer concentration at a downstream sampling point in tracer studies.

which is sketched in Figure 4.18. By fitting the measured concentration curve to this equation, V and D_t can be measured simultaneously. In practice, there are some problems of interpretation. For example, the velocity measured in this way for real flows would represent some sort of average between the tracer introduction point and the sampling point. Furthermore, there is usually some uncertainty about the path pursued by the fluid in moving from the tracer introduction point to the sampling point. A correction for the decay of the tracer is frequently required.

Grjotheim and coworkers[32] have made extensive use of radiotracers in their studies of aluminum reduction cells. In addition to velocity and diffusivity measurements, these workers have used the technique to measure current efficiency, an application further developed by Naeuman and Torklep.[33] Figure 4.19 gives the velocity vectors measured on metal and electrolyte using the radiotracer technique. These workers reported a turbulent diffusivity of 5 cm^2/s for the metal.

Szekely[35] used a radioactive gold tracer to measure the turbulent diffusivity within the metal in an open-hearth furnace, obtaining values of the order of 20 cm^2/s.

Tracer dispersion studies on a 50-ton ASEA-SKF furnace were reported by

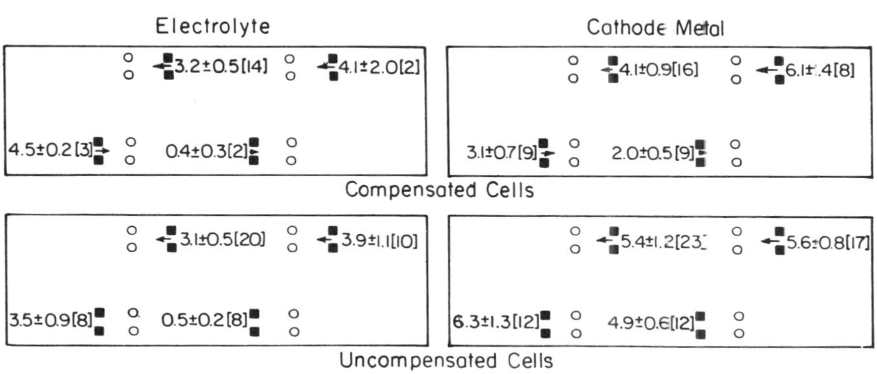

FIG. 4.19. Velocities (cm/s) measured in aluminum cells by Berge et al. using the radiotracer technique.[34]

Nakanishi et al.[36] These investigators reported eddy diffusivities of about 100–300 cm^2/s.

Szekely and Chang[37] reported the results of some measurements made at British Steel Corporation wherein radioactive gold was used to measure the velocity of an inductively stirred melt.

It should be mentioned that tracers are used frequently to study the residence time distribution of fluid elements moving through a vessel. Such studies are employed on continuous flow systems with the tracer injected at the fluid inlet and its presence measured as a function of time at the fluid outlet by an appropriate sensor or by sampling.

Related to this kind of study are investigations of mixing in batch systems. Here the tracer is injected at some representative point and its concentration measured as a function of time at one or more other points. It should be noted, however, that mixing is a consequence of both diffusion (eddy diffusion if the flow is turbulent) and bulk flow; it is therefore sometimes difficult to interpret the results of mixing experiments of this nature, although the practical value of such studies is high.

4.3. TEMPERATURE MEASUREMENT

4.3.1. Thermocouples

The conventional thermocouple finds widespread application in pyrometallurgical applications. The literature on this device is extensive, and it will not be discussed here except by way of Table 4.3.

The intermittent use of a thermocouple is described by Raeger and Olsen[39] and by Giedt and Chambers.[40] In this application the rate of increase of the thermocouple tip temperature, after insertion into the environment whose temperature is to be measured, is monitored (by following the thermocouple emf). The thermocouple is withdrawn as it approaches its maximum safe temperature. The rate of increase of the tip temperature is a function of the environment temperature. Unfortunately the rate is also a function of the heat transfer coefficient, requiring a prior knowledge of this quantity or the use of two thermocouples of the same geometry but different masses.

At temperatures at which radiation is significant, a thermocouple placed in an optically thin enclosure will yield a temperature that is some composite of the surface temperatures of the walls, rather than of the fluid in the enclosure. This difficulty can be avoided by use of the suction pyrometer wherein the thermocouple tip is screened by radiation shields and the fluid flow past the thermocouple tip is increased by sucking the gas past the tip. This has the effect of causing convective heat transport from the fluid.

4.3.2. Pneumatic and Venturi Pyrometers

The pressure differential developed by an orifice plate or a venturi through which a gas is flowing is dependent on geometry, mass flow rate, and the density of the gas. The density in turn is a function of temperature (together with

TABLE 4.3. Characteristics of Common Thermocouples

Type	Normal Temperature Limit (°C)	Approximate CMF at Normal Temperature Limit (mV)	Short-Term Temperature Limit (°C)	Resistance to Atmosphere		Melting Point of Lowest MP Limb (°C)	Sensitive to	Compensating Cable	Notes
				Oxidizing	Reducing				
Chromel/Alumel (nickel–chromium/ nickel–aluminium)	1100	45	1300	Good	Poor	1430	Cyanide fumes, sulfur	Chromel/ Alumel	
Palladium/platinum, gold/palladium (Platinel II, Pallador II)	1200	49		Good	Poor	1460		Chromel/ Alumel	Precious metal substituted for Chromel/Alumel similar electrical output
Platinum/platinum: 13% rhodium	1400	16	1650	Very good	Very poor	1770	Silica, metallic vapors	Copper copper: 2% nickel	
Platinum: 5% rhodium, platinum: 20% rhodium	1500	9	1700	Very good	Very poor	1805	Silica, metallic vapors	Copper	
Iridium: 40% rhodium/ iridium	2000	11	2100	Fair	Good	2250		Copper/ aluminium	Most stable of Ir/Rh combinations
Tungsten: 5% rhenium/ tungsten	2600	39		Very poor		>3000		Usually copper	Thoria insulation for high temperatures

SOURCE: From Roberts et al.[38]

molecular weight and pressure). This is the principle of operation of pneumatic and venturi pyrometers, wherein the gas flows (without significant cooling) through an orifice plate or venturi, yielding the high temperature–pressure differential, and then through a cooling section and through a second orifice plate, or venturi, yielding the mass flow rate.

4.3.3. Optical Pyrometers

The optical pyrometer is a widely used tool for temperature measurement, both in the laboratory and in actual metal processing operations. The conventional instrument only yields accurate results if the emissivity of the surface whose temperature is being measured is close to unity. For real surfaces, with emissivities below unity, the optical pyrometer will usually indicate a temperature that is lower than the actual temperature. However, this is not always the case. For example, a cold body of high reflectivity (low emissivity) located in a hot furnace may reflect sufficient radiation impinging on it to give an artificially *high* temperature as measured by the optical pyrometer.

Ratio, or two-color, pyrometers are claimed to overcome the emissivity problem. The principle of their operation can be discerned from Figure 4.20. A gray surface is one whose monochromatic emmisivity is independent of wavelength. Consider the two wave bands (colors) indicated by the vertical broken lines in Figure 4.20. Clearly, at higher temperatures, the radiant power in these wave bands will be higher for both wave bands; however, the power in the short-wavelength wave band increases much more rapidly than that in the long-wavelength wave band with increasing temperature. Clearly, then, the ratio of the powers provides a measure of temperature and this ratio is independent of emissivity, provided the surface is gray. Unfortunately, many surfaces are not gray, and there is no guarantee, for any particular application, that the two-color pyrometer will give more accurate results than the conventional optical pyrometer. A readable discussion of this point is contained in Roberts et al.[38]

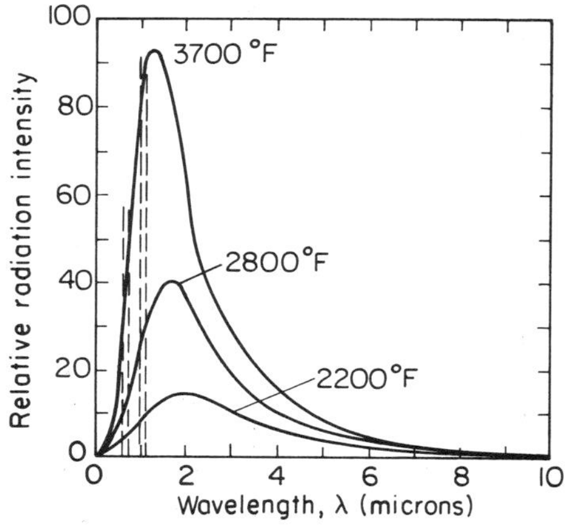

FIG. 4.20. Monochromatic emissive power of a gray surface as a function of wavelength and temperature.

Radiation pyrometers, wherein the radiant energy from a hot source is measured thermoelectrically or photoelectrically, are also in common use and are discussed at some length in Roberts et al.[38] As with optical pyrometers, care must be taken to avoid obstructions and hot objects in the line of sight from the source whose temperature is to be measured. Again, this pyrometer is sensitive to the emissivity of the source. This sensitivity is somewhat less for the partial radiation pyrometer, which is sensitive only to the near infrared region of the spectrum. In this region (see Figure 4.20) the monochromatic emissive power is much more sensitive to temperature than is the fourth-power dependence of the total emissive power (measured by the conventional total radiation pyrometer). Consequently, an error in emissivity corresponds to a much smaller error in temperature.

Some methods are available for overcoming the emissivity problem in optical and radiation pyrometry, for example, forming an insulated cavity around the object and allowing thermal equilibrium to be reached. Radiation streaming through a small hole in the cavity is then black body radiation at the temperature of the cavity.

Alternatively, a hole can be drilled in the object such as to approximate a black-body cavity. Iuchi[41] has recently described a device for simultaneous measurement of temperature and emissivity and the application of this device to measurements of the output from a strip-coating machine. The principle is illustrated in Figure 4.21. The object whose temperature is to be measured (e.g., the steel strip) forms the bottom of a cylindrical cavity, the curved wall of which is a specularly coated surface held at a much lower temperature by cooling water and air. With no top plate present, nearly all the radiation leaving the object streams from the top of the cavity, and the emissive power at the cavity top is therefore

$$E_1 = \varepsilon \sigma T^4$$

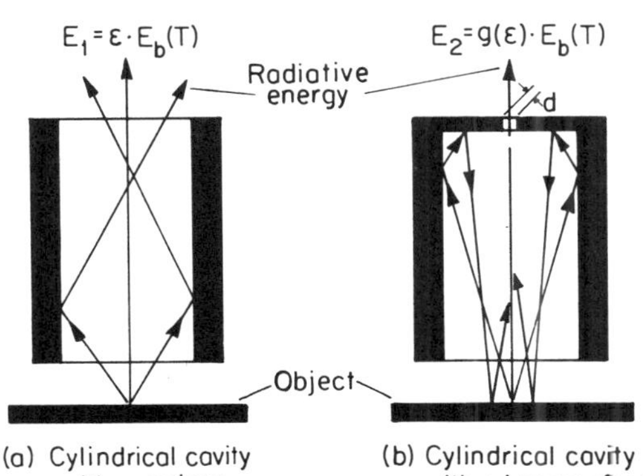

FIG. 4.21. Principle of Iuchi's device for simultaneous measurement of temperature and emissivity.[41]

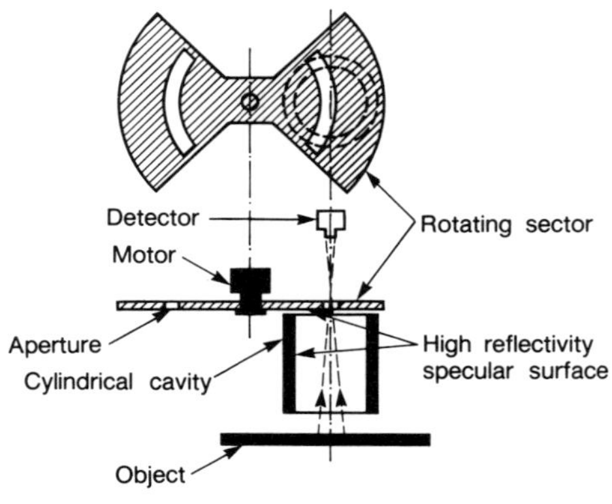

FIG. 4.22. Mechanical details of Iuchi's apparatus.[41]

where ε is the emissivity of the object, and
T is the temperature of the object.

Next, a cover with a small aperture is positioned on top of the cylinder. The emissive power of the radiation streaming from the cavity is now much higher because of reflection within the cavity:

$$E_2 = g(\varepsilon)\sigma T^4$$

where the function

$$g(\varepsilon) = \frac{(\alpha + 1)\varepsilon}{(\varepsilon + \alpha)}$$

and α is dependent on the geometry and reflectivity of the coated surfaces but not on the emissivity of the object. Knowing α, which can be obtained from calibration experiments, the values of T and ε can be determined simultaneously. The details of the apparatus built by Iuchi appear in Figure 4.22.

An interesting recent development is optical fiber thermometry. In this technique a black-body sensor is formed at the end of an optical fiber, the opposite end of which is attached to a photodetector. The output of the photodetector is a measure of the temperature of the black-body sensor. Typically the optical fiber is a sapphire one with a high resistance to corrosive high-temperature environments. The device offers high precision and fast response to temperature changes but is limited to temperatures above approximately 300°C.

4.3.4. Other Temperature Measurement Techniques

Numerous temperature-sensitive paints, chalks, stickers, and plugs are commercially available for convenient measurement of surface temperature. Suzuki and Ohno[42] have recently reviewed the trends in temperature measurement technology in the steel industry.

Finally, we shall mention the use of infrared photography and thermal scanning to obtain semiquantitative or quantitative temperature profiles on the surface of objects. The latter is likely to see widespread application in the future because of the ease of processing and storing images using video recorders.

4.4. MEASUREMENTS OF REACTION RATES

Homogeneous reactions are seldom encountered in process metallurgy, and this discussion is therefore limited to heterogeneous reactions, in particular, to the measurement of the rate of reaction between a solid and a gas or liquid. The objective of such measurements is to obtain a quantitative relationship between the reaction rate and such independent variables as fluid reactant concentration and temperature. The concentration and temperature are fairly well defined if experiments are carried out on a single entity of solid (e.g., an iron ore pellet or a single lump of ore). This is in contrast to experiments on a collection of entities (e.g., a packed bed), where substantial variations of concentration, and perhaps temperature, can be expected. Consequently, it makes sense to carry out laboratory experiments on single pellets or lumps of solid.

4.4.1. The Gravimetric Apparatus

For gas–solid reactions the most widely used equipment for measuring reaction rate is the gravimetric apparatus illustrated in Figure 4.23. The lump of solid is

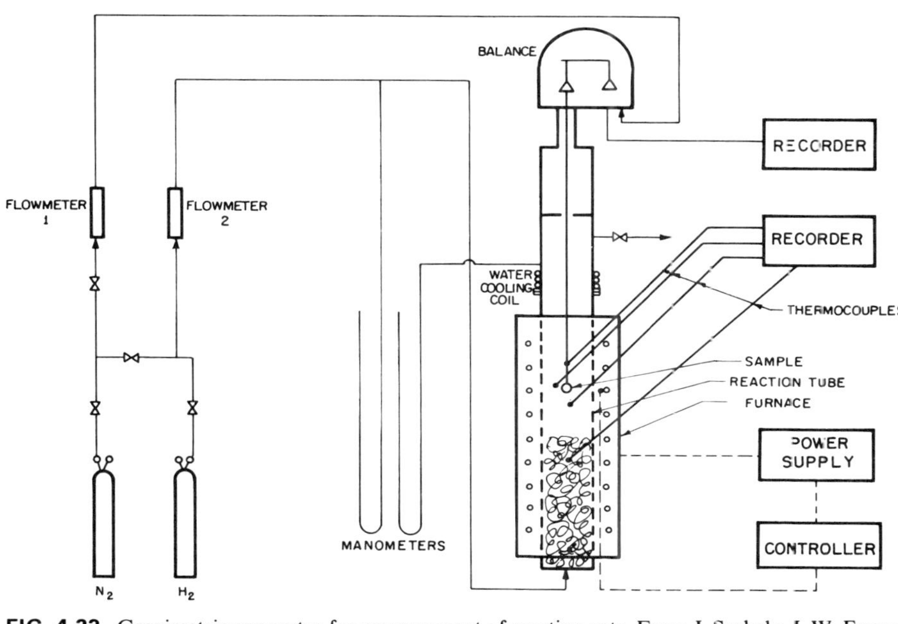

FIG. 4.23. Gravimetric apparatus for measurement of reaction rate. From J. Szekely, J. W. Evans, and H. Y. Sohn, *Gas–Solid Reactions*, Academic Press, New York, 1976.

suspended from one arm of a balance, usually an electronic recording balance, although frequently quartz spring balances have been used. The solid is located within a reaction tube, which in turn is located within a furnace that is usually held at constant temperature by means of an electronic controller. The gaseous reactant is passed through the reaction tube at a measured flow rate, usually via some type of preheater to bring the gas up to the furnace temperature. The progress of reaction is followed by determining the weight change resulting from reaction.

Experimental Design

It is important that the reactor be operated in a differential manner, that is, with only a small change in the gaseous reactant concentration due to reaction; in this way the concentration of gaseous reactant is well defined. This implies that the gaseous reactant should be passed through the reaction tube at a rate that is high compared to the reaction rate. A second requirement for the flow rate occurs due to considerations of external mass transfer. Usually it is preferable to carry out the reaction so that mass transfer external to the solid has little or no influence on the progress of reaction, which again implies a high gas flow rate. The technique most frequently used for establishing the necessary gas flow rate is carrying out measurements at higher and higher flow rates until the reaction rates are not significantly affected by the flow rates.

A second important consideration arises from the relative importance of pore diffusion or the heterogeneous reaction itself. This is illustrated in Figure 4.24. At high temperatures or large lump sizes, the rate-controlling step is the diffusion of gaseous reactant within the pores of the solid. Conversely, the chemical step tends to be controlling at lower temperatures or small lump sizes. If it is necessary to measure rates of the chemically controlled reaction at intermediate temperatures, difficulties may be encountered in using the gravimetric technique, since there are limits on how small a lump can be made without loss of accuracy in weight measurement. One technique is to do experiments on small lumps and extrapolate the results to zero size.[43] An alternative approach is to abandon the gravimetric technique and follow the

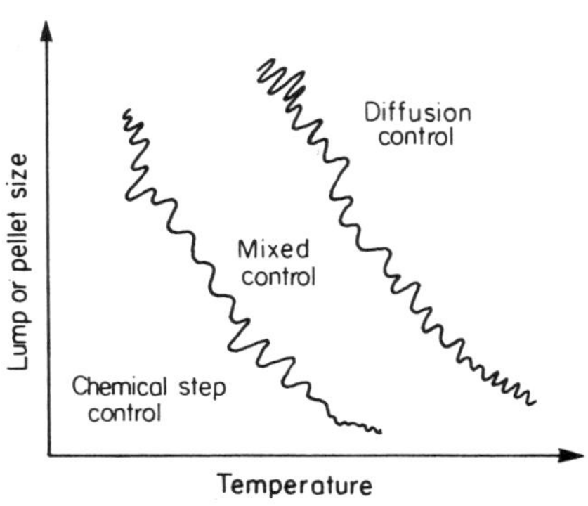

FIG. 4.24. Schematic diagram of the regimes of control in gas–solid reactions. From J. Szekely, J. W. Evans, and H. Y. Sohn, *Gas–Solid Reactions*, Academic Press, New York, 1976.

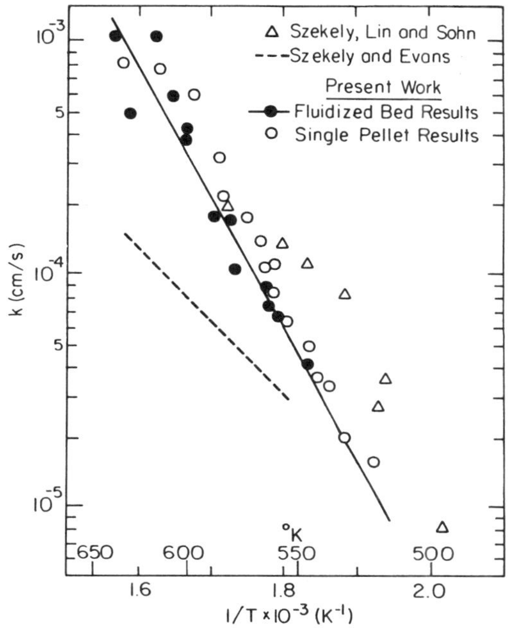

FIG. 4.25. Comparison of rate measurements for nickel oxide reduction made using a laboratory fluidized bed with those made using the gravimetric technique. From J. W. Evans, and S. Song, *Proceedings of the Fourth Inter-American Conference on Materials Technology*, ASME, Caracas, Venezuela 1975.

progress of reaction in some other way. For example, Evans et al.[44] used a small fluidized bed to study the reaction of nickel oxide particles with hydrogen. The reaction was carried out under differential conditions (so that the gaseous reactant concentration was well defined) and the progress of reaction was monitored by measurement of the small amount of water vapor leaving the reactor. Figure 4.25 is a comparison of results obtained in this way with results using the gravimetric technique. A similar technique has been used by Koo and Evans[45] for the measurement of the rate of reduction of Venezuelan iron ores.

4.4.2. The Torsion Technique

A novel technique for measurement of reaction rate has been used by Evans and Haase[46] and is illustrated in Figures 4.26 and 4.27. Lumps of solid reactant (e.g., iron oxide) are mounted on opposite arms and opposite faces of a torsion cell that is made of a nonreactive material. The torsion cell hangs by a very fine suspension wire within a furnace, and its deflection can be measured by means of a light beam. Consider what happens on introducing a reactive gas (e.g., hydrogen) into the environment surrounding the cell. Gas molecules bombard the cell and for the most part are reflected unchanged. However, a fraction of the hydrogen molecules striking the iron oxide faces are converted to water vapor

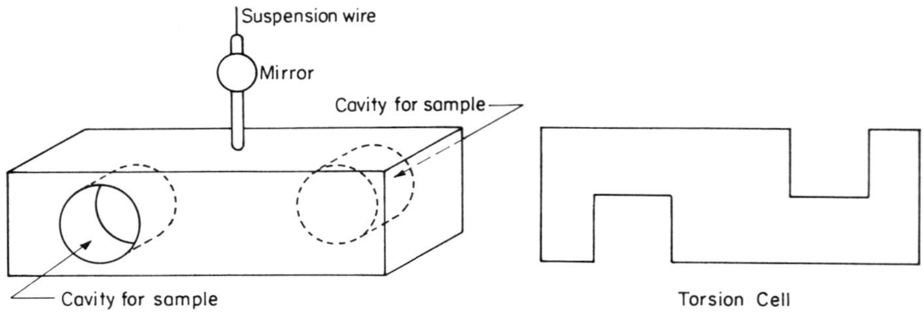

FIG. 4.26. The torsion cell for measurement of rates of gas–solid reactions.

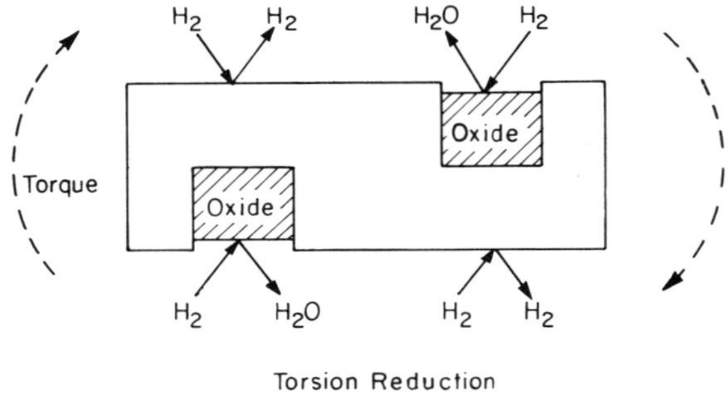

FIG. 4.27. Principle of operation of the torsion technique in oxide reduction reactors.

before leaving. For these molecules the exchange of momentum on collision is different from that of a normal collision due to the different molecular weight of the striking and leaving molecule. The net effect is torque on the cell, which can be measured by the deflection of the cell, the torque being proportional to reaction rate. This technique is experimentally more difficult than the gravimetric technique and is unlikely to replace it.

4.4.3. Liquid–Solid Reactions

The gravimetric technique is not suitable for liquid–solid reactions, since the buoyancy and drag forces on a lump of solid make accurate weight determination difficult. The standard technique is to immerse a few lumps of solid in an agitated, thermostatted excess of liquid reactant and to monitor the progress of reaction by analysis of the liquid for reaction product. It is usual to do several runs at different agitator speeds until an agitator speed is reached at which the measured rates are insensitive to the agitator speed. However, caution must be exercised, since at agitator speeds at which the solids become suspended, the mass transfer coefficient is no longer sensitive to agitator speed. Consequently, external mass transfer may exert an influence on the progress of reaction even though the rates are found insensitive to agitator speed in the experiments.

In Section 4.4.1, the influence of lump size and temperature on the rate-controlling step in gas–solid reactions was discussed. Similar considerations apply to liquid–solid reactions. In reactions encountered in hydrometallurgy, two other facts need to be borne in mind. First, hydrometallurgical ores are frequently much more heterogeneous than the solids encountered in pyrometallurgy, making more difficult the obtainment of representative samples and the interpretation of experimental measurements. Second, in many hydrometallurgical reactions a third, gaseous phase is involved, for example, in the oxidative leaching of sulfide ores using oxygen as the oxidizing agent. Frequently autoclaves and specialized techniques are required for the experimental study of such reactions.

4.5. CONCLUDING REMARKS

This survey of experimental techniques has necessarily been brief, even though confined to measurements of velocity, temperature, and reaction rate. These techniques are currently in a rapid state of development, particularly the measurement of velocities. One important development is the linking of the measurement device to micro- or minicomputers. The laser–Doppler velocimeter is an example of this. The device is capable of measuring velocities several thousand or even hundreds of thousands of times per second; only by using a computer for data acquisition and processing is this capability fully exploited. Another example is provided by the optical fiber thermometer; in the commercial embodiment of this device a microcomputer is used to relate the photodetector output to temperature, to store temperature measurements, and to process the data. Finally, the robotization of experimental measurement is becoming common. The work of Taberlet[10] described in Section 4.1.1 is an example. In that investigation the measurement of velocities by hot-film velocimetry and the location of the thermal wake was carried out, throughout the flow, entirely automatically under the control of a minicomputer. Such robotization generally allows much faster and less expensive acquisition of data from physical models, and its more widespread use, both on models and physical plants, can be expected.

REFERENCES

1. J. W. Evans, *Min. Sci. Eng.*, **11**, 207 (1979).
2. J. Szekely and R. T. Yadoya, *Met. Trans.*, **3**, 2673 (1972).
3. R. Mutharason and D. Apelian, Paper 15a, AIChE Seventy-second Annual Meeting, San Francisco, November 1979.
4. E. D. Tarapore, and J. W. Evans, *Met. Trans.*, **7B**, 343 (1976).
5. W. Merzkirch, *Flow Visualization*, Academic Press, New York, 1974

6. A. T. Hjelmfelt and L. F. Mockros, *Appl. Sci. Res.*, **16**, 149 (1966).
7. L. J. Saunders and P. Lawrence, in *Fluid Dynamic Measurement*, D. Cockrell, ed., Leicester University Press, Leicester, England, 1972, p. 125.
8. R. W. Johnson and A. M. Dhanak, in *Fluid Dynamic Measurement*, D. Cockrell, ed., Leicester University Press, Leicester, England, 1972, p. 131.
9. T. W. Davies and M. A. Patrick, in *Fluid Dynamic Measurement*, D. Cockrell, ed., Leicester University Press, Leicester, England, 1972, p. 152.
10. E. Taberlet, Le Brassage Turbulent dans les Fours a Induction, doctoral thesis, Institut National Polytechnique de Grenoble, France, 1984.
11. M. Sajben, *Rev. Sci.*, **39**, 945 (1965).
12. C. A. Sleicher and G. B. Lim, Proceedings of the Symposium on Turbulence in Liquids, Rolla, 1973, Univ. of Missauri-Rolla (1975), p. 1.
13. P. Cremer and A. Alemany, *J. Mechan. Appl.*, **15**, 37 (1981).
14. A. Murthy and J. Szekely, *Met. Trans.*, **14B**, 499 (1983).
15. J. Janalik, in *Fluid Dynamic Measurement*, D. Cockrell, ed., Leicester University Press, Leicester, England, 1972, p. 175.
16. G. Q. Martin and L. N. Johanson, *AIChE J.* **11**, 29 (1965).
17. R. Ricou and C. Vives, *Int. J. Heat Mass Transfer*, **25**, 1579 (1982).
18. H.-C. Lee, J. W. Evans, and C. Vives, *Met. Trans.*, **15B**, 734 (1984).
19. F. Durst, A. Melling, and J. H. Whitelaw, *Principles and Practices of Laser–Doppler Anermometry*, Academic Press, New York, 1976.
20. R. L. Humphrey, *Liquid Turbulence and Its Measurements*, Science Press, Princeton, NJ, 1977, p. 103.
21. D. J. Moore and J. C. R. Hunt, Proceedings of the Third Bar-Sheva Symposium on MMD Flows and Turbulence, Beersheva, Israel, 1981.
22. A. Moros, Drag Aneometry for Measuring Velocities in Induction Furnaces, Int. Seminar on Refining and Alloying of Liquid Aluminum and Ferro-Alloys, Norwegian Institute of Technology, Trondheim, Norway, August, 1985.
23. D. R. Sadoway and J. Szekely, *Met. Trans.*, **11B**, 334 (1980).
24. E. D. Tarapore, J. W. Evans, and J. Langfeldt, *Met. Trans.*, **8B**, 179 (1977).
25. J. Szekely, C. W. Chang, and W. E. Johnson, *Met. Trans.*, **8B**, 514 (1977).
26. A. A. Bustos, G. G. Richards, N. B. Gray, and J. K. Brimacombe, *Met. Trans.*, **15B**, 77 (1984).
27. M. Kase, M. Sugata, K. Yamaguchi, and M. Nakagome, *Trans. ISI Japan*, **22**, 811 (1982).
28. K. Yamaguchi, M. Nakagome, M. Sugata, H. Nishikawa, and Y. Abe, *Trans. ISI Japan*, **25**, 371 (1985).
29. J. Szekely, C. W. Chang, and R. E. Ryan, *Met. Trans.*, **8B**, 333 (1977).
30. N. El-Kaddah, J. Szekely, and G. Carlsson, *Met. Trans.*, **15B**, 633 (1984).
31. R. W. Johnson, in *Light Metals 1978*, Metallurgical Soc. of AIME, New York, 1978, p. 45.
32. K. Grjotheim et al, *Met. Trans.*, **1**, 3133 (1970); **2**, 199 (1971); **4**, 1945 (1973).
33. R. Næuman and K. Torklep, Determination of Metal Inventory and Current Efficiency in Commercial Aluminum Reduction Cells, Extended Abstracts, Thirtieth Annual Meeting of the International Society of Electrochemistry, Trondheim, Norway, 1979.
34. B. Berge, K. Grjotheim, C. Krohn, R. Næumann, and K. Tørklep, *Met. Trans.*, **4**, 1945 (1973).
35. J. Szekely, *J. ISI*, **202**, 505 (1964).
36. K. Nakanishi, J. Szekely, T. Fujii, Y. Mihara, and S. Iwaoka, *Met. Trans.*, **6B**, 111 (1975).
37. J. Szekely and C. W. Chang, *Ironmaking and Steelmaking*, **4**, 196 (1977).
38. F. Roberts, R. F. Taylor, and T. R. Jenkins, eds., *High Temperature Chemical Reaction Engineering*, Institution of Chemical Engineers, London, 1971.

39. American Institute of Physics, *Temperature: Its Measurement and Control in Science and Industry*, vol. 3, Reinhold, New York, 1962, p. 401.

40. W. H. Giedt and J. T. Chambers, *Trans. ASME J. Heat Transfer Ser. C*, **87**, 397 (1965).

41. T. Iuchi, *Trans. ISI Japan*, **19**, 474 (1979).

42. H. Suzuki and J. Ohno, *Trans. ISI Japan*, **19**, 440 (1979).

43. J. Szekely, C. I. Lin, and H. Y. Sohn, Chem. Eng. Sci., **28**, (1973) 1975.

44. J. W. Evans, S. Song, and C. E. Leon-Sucre, *Met. Trans.*, **7B**, 55 (1976).

45. C.-H. Koo and J. W. Evans, *Trans. ISI Japan*, **10**, 95 (1979).

46. J. W. Evans and K. Haase, *High Temp. Sci.*, **8**, 167 (1976).

5 MODELING OF FLUID – SOLID REACTORS

This chapter outlines suggested procedures for the modeling of reactors in which solid reacts with a gas or a liquid. Examples of such reactors are direct reduction units (where a packed, moving, or fluidized bed of iron ore is contacted with a reducing gas) and leaching columns (where, e.g., a copper oxide ore is leached with sulfuric acid).

The objective of such modeling is usually to predict the performance of the reactor, that is, to predict the overall rate of conversion of the solid contained in the reactor. The first task is to arrive at a quantitative description of the rate of reaction of an individual entity of solid (particle, ore lump, or pellet) contained in the reactor for a *specified* environment surrounding this entity (specified fluid composition, temperature, etc.). This is only the first task, however, since in most reactors the environment surrounding one particle is affected by the presence of the other particles, most notably by the reaction of the other particles. A second task is, therefore, to apply transport and conservation equations to describe the environment in the reactor, which is usually a function of time and position.

5.1. MODELING OF THE REACTION OF AN INDIVIDUAL ENTITY OF SOLID

Two main classes of models are available for describing the reaction of a lump or pellet of solid with a fluid. The first class of model is the shrinking core model, or similar models, which can describe adequately the reaction of an initially nonporous lump of solid with a fluid. This model has been in existence for a few decades and is illustrated in Figure 5.1.

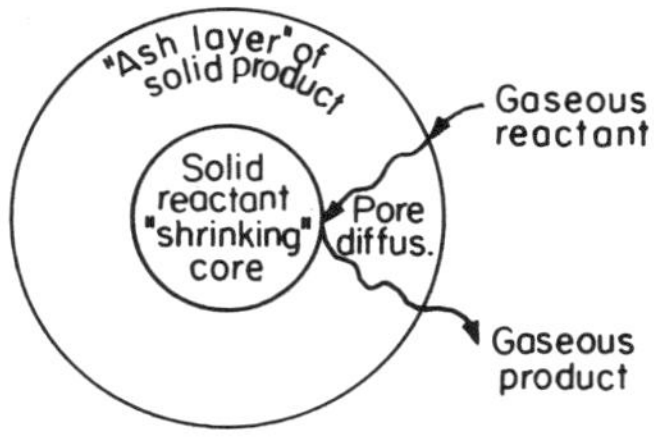

FIG. 5.1. Typical morphology of a nonporous solid that has partially reacted with a fluid.

We will now describe the development of the model. Consider the reaction

$$\underset{\text{(fluid reactant)}}{\mathrm{A}} + \underset{\text{(solid reactant)}}{b\mathrm{B}} = \underset{\text{(fluid product)}}{\mathrm{C}} + \underset{\text{(solid product)}}{d\mathrm{D}} \tag{1}$$

and let us suppose that B is initially nonporous, while the solid product D is porous. This would be the case for most ores in their native state (such ores typically have very low porosity) reacting to give a porous mass of solid reaction product or residue. The unreacted core of B shrinks as the reaction proceeds. Because B is nonporous, the fluid reactant cannot penetrate it, and reaction is confined to the interface between the core and the porous layer of D. The latter is sometimes called the ash layer. The fluid reactant has to diffuse through the ash layer in order to reach the reaction interface, and subsequently the fluid product diffuses out.

Let us first make some simplifying assumptions. These assumptions are to some extent unrealistic, but our intention here is to be informative rather than precise in the development of a shrinking core model. More sophisticated versions of the model avoid some of these simplifications. It will be assumed that the lump of solid is spherical and that the shrinking core is spherical and concentric with the lump. Furthermore, it will be assumed that the reaction is irreversible.

It is clear that a spherical coordinate system is appropriate for this model, as depicted in Figure 5.2. It is also appropriate to select a spherical volume element (shell) of differential thickness for a mass balance, and this too is depicted in Figure 5.2. The mass balance is on the fluid reactant A and is formulated as follows.

Rate of diffusion of A into the shell through the outside of the shell (at $r_s + dr$) is

$$\left(D_{\mathrm{Aeff}} \frac{\partial C_{\mathrm{A}}}{\partial r} 4\pi r^2\right)_{r_s + dr} \tag{2}$$

Here we have used Fick's first law to obtain the flux $D_{\mathrm{Aeff}}(\partial C_{\mathrm{A}}/\partial r)$ into the shell and then multiplied by the surface area of the shell. D_{Aeff} is the effective diffusivity of the fluid within the porous ash layer, and C_{A} is the concentration of A within the pores. The subscript on the parentheses means that everything

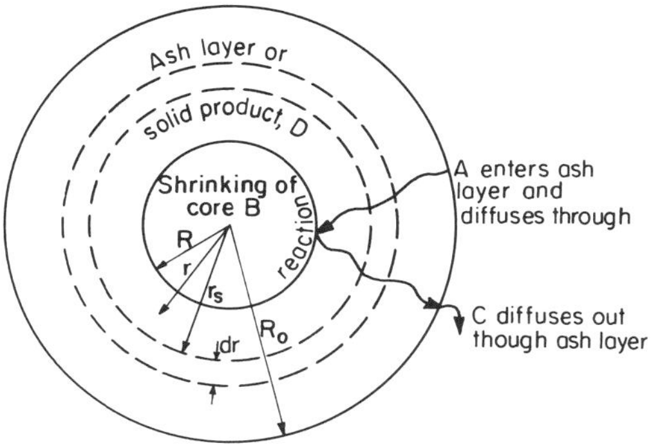

FIG. 5.2. Shrinking core model for gas–solid reactions.

within the parentheses (including r^2) is to be evaluated at $r_s + dr$. We have assumed here that diffusion is the sole mechanism of mass transport within the pores (i.e., that flow resulting from any total pressure gradients and transport due to bulk motion of the fluid in the pores is negligible).

Similarly, the rate of diffusion of A out of the shell at r_s is

$$\left(D_{\text{Aeff}} \frac{\partial C_A}{\partial r} 4\pi r^2\right)_{r_s} \tag{3}$$

The rate of accumulation of A in the shell is

$$\left[\frac{\partial}{\partial t}(4\pi r^2\, dr\, \varepsilon C_A)\right]_{r_s} = 4\pi r_s^2 \varepsilon\, dr \left(\frac{\partial C_A}{\partial t}\right)_{r_s} \tag{4}$$

where the porosity ε has been introduced because only a fraction ε of the total shell volume $4\pi r_s^2\, dr$ is pore space. A is neither destroyed nor created in the pores, since reaction is taking place only at $r = R$. Hence, we can write

$$\text{input} = \text{output} + \text{accumulation}$$

and writing this in mathematical symbols we get (after some cancellation and rearranging)

$$\frac{(D_{\text{Aeff}} r^2\, \partial C_A/\partial t)_{r_s+dr} - (D_{\text{Aeff}} r^2\, \partial C_A/\partial t)_{r_s}}{dr} = r_s^2 \varepsilon \left(\frac{\partial C_A}{\partial t}\right)_{r_s} \tag{5}$$

Recognizing that dr is differential, we can write the left-hand side as the derivative of the quantity in parentheses evaluated at r_s, or

$$\frac{1}{r_s^2}\left[\frac{\partial}{\partial r}\left(D_{\text{Aeff}} r^2 \frac{\partial C_A}{\partial r}\right)\right]_{r_s} = \varepsilon \left(\frac{\partial C_A}{\partial t}\right)_{r_s} \tag{6}$$

However, r_s was chosen arbitrarily in the range $R < r_s < R_0$, and so this equation applies for any value of r and there results

$$\frac{1}{r^2}\frac{\partial}{\partial r}\left(D_{\text{Aeff}} r^2 \frac{\partial C_A}{\partial r}\right) = \varepsilon \frac{\partial C_A}{\partial t} \tag{7}$$

At his point in the development of the shrinking core model, one makes the quasi-steady-state approximation and assumes that the right-hand side of Eq. (7) is approximately zero. This is equivalent to assuming that the relaxation time for the establishment of a steady concentration profile in the ash layer is small compared to the time to react the solid. The assumption is a valid one for gas–solid reactions of practical interest but is questionable for some fluid–solid reactions (e.g., the leaching of lean ores).

With the right-hand side of Eq. (7) set equal to zero, it can readily be integrated to obtain

$$D_{\text{Aeff}} \frac{\partial C_A}{\partial r} = \frac{C_1}{r^2} \tag{8}$$

where C_1 is an integration constant. If the solid D is macroscopically uniform, D_{Aeff} will be independent of r and a second integration can be carried out:

$$D_{\text{Aeff}} C_A = -\frac{C_1}{r} + C_2 \tag{9}$$

The integration constants can be evaluated from the boundary conditions, which are

$$\begin{aligned} &BC1 \quad \text{at} \quad r = R_0 \qquad C_A = C_{AO} \\ &BC2 \quad \text{at} \quad r = R \qquad D_{\text{Aeff}} \frac{\partial C_A}{\partial r} = kC_A \end{aligned} \tag{10}$$

Let us assume here that the mass transfer step from the bulk of the fluid into the external surface of the solid represents negligible resistance to the progress to reaction. C_{AO} is then the bulk concentration of the fluid reactant, and this justifies $BC1$. The second boundary condition arises from equating the flux of A to the reaction interface to its rate of reaction there. That rate is assumed here to be given by a first-order rate expression with k as the chemical rate constant. Equations (8)–(10) contain the unknowns C_A, C_1, and C_2, and simple algebra leads us to an equation for C_A as a function of r (i.e., the concentration profile) with R_0, R, C_{AO}, k, and D_{Aeff} as parameters. This result is not, however, of much practical interest in itself. Of more importance is the rate of reaction of the solid or, expressed another way, the rate at which the core is shrinking. This is obtained from

$$-\frac{dR}{dt}\rho_s = bD_{\text{Aeff}}\frac{\partial C_A}{\partial r} \tag{11}$$

Equation (11) is obtained by equating the rate of reaction of B at the interface to b times the rate of arrival of A at the interface. The molar density of the solid reactant (moles of B per unit volume of solid) is ρ_s. This ordinary differential equation has the initial condition

$$IC \qquad R = R_0 \quad \text{at} \quad t = 0 \tag{12}$$

The right-hand side of this ordinary differential equation is obtained (as a function of R_0, R, C_{Ao}, k, and D_{Aeff}) from the concentration profile we have derived, and the equation can then be integrated analytically and the initial condition used to obtain the integration constant. These straightforward mathematical manipulations are left as an exercise for the reader. The final result is

$$\frac{R_o k}{6D_{\text{Aeff}}}[3 - 3(1 - X)^{2/3} - 2X] + [1 - (1 - X)^{1/3}] = \frac{kC_{Ao}b}{R_o\rho_s}t \tag{13}$$

where X is the extent of reaction-fraction of B that has been converted into $D = 1 - (R/R_o)^3$. A sketch of this result appears in Figure 5.3.

EXAMPLE 1

Laboratory experiments have determined a value of $k = 0.1$ cm/s at 800 K for the reduction of nickel oxide by hydrogen. In the same experiments an effective diffusion coefficent of 0.01 cm^2/s was measured for the diffusion of hydrogen and water vapor within the porous nickel produced by reduction. Estimate the time to achieve 90% reduction of 2 mm diameter dense nickel oxide spheres at this

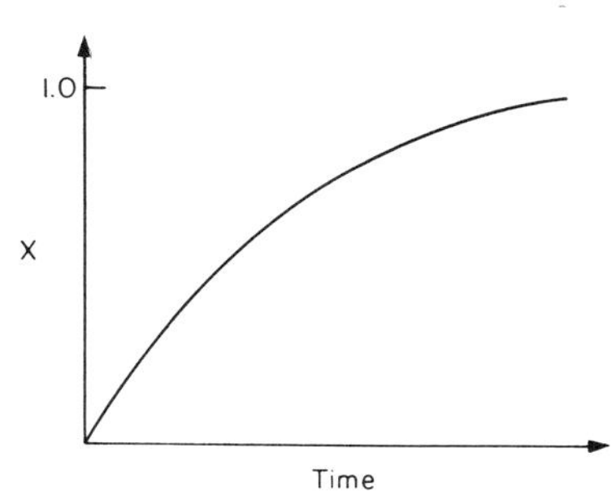

FIG. 5.3. Typical plot of extent of reaction versus time.

temperature in 1 atm of hydrogen. Additional data are

$$\text{density of NiO} = 6.7 \text{ g/cm}^3$$

$$\text{molecular weight} = 74.7 \quad (\text{i.e., } \rho_s = 0.0897 \text{ mole/cm}^3)$$

Solution. From the ideal gas law,

$$C_{Ao} = P/RT = 1.52 \times 10^{-5} \text{ mole/cm}^3$$

For $X = 0.9$, Eq. (13) becomes

$$\frac{0.001 \text{ m} \times 10^{-3} \text{ m/s}}{6 \times 10^{-6} \text{ m}^2\text{/s}}(3 - 0.6463 - 1.8) + (1 - 0.4641) = \frac{10^{-3} \text{ m/s} \times 18.5 \text{ mole/m}^3\, t}{0.001 \text{ m} \times 89.7 \times 10^3 \text{ mole/m}^3}$$

since for this reaction $b = 1$, $t = 3045 \text{ s} = 51 \text{ min}$.

The model predicts that the rate of reaction declines as reaction proceeds, which is the usual experimental observation. Some other predictions of the model are interesting. The time for complete reaction is given by

$$t_{X=1} = \frac{\rho_s R_o}{bC_{AD}}\left(\frac{R_o}{6D_{Aeff}} + \frac{1}{k}\right) \tag{14}$$

From this it is clear that the time to completely react a solid will be proportional to $(R_o)^2$ for $R_o k/6D_{Aeff} \gg 1$ (which corresponds to pore diffusion control), but to the first power of R_o for $R_o k/6D_{Aeff} \ll 1$ (which corresponds to chemical step control).

Many investigators have described variants of the shrinking core model, and reference can be made to the literature.[1] It should be noted that, although the above development concerns a spherical lump of solid, shrinking core models are available for solids of other simple shapes. Under some circumstances the model may be applied to a solid that is initially porous; however, for initially porous solids it is better to employ the alternative class of models, of which the grain model is an example. This model is sketched in Figure 5.4.

The initially porous solid is considered to be made up of nonporous grains. In Figure 5.4 these grains are depicted as spheres, but this is not a necessary assumption. It is seen that the grain model differs from the shrinking core model in that the reaction interface of the latter is smeared out in the grain model. This is because, for an initially porous solid, the fluid reactant molecules can penetrate between the grains of the solid reactant before they react.

The differential equation describing the diffusion of the fluid reactant within the solid now contains a term accounting for the local rate of reaction within the solid. (It will be recalled that for the shrinking core model the reaction rate

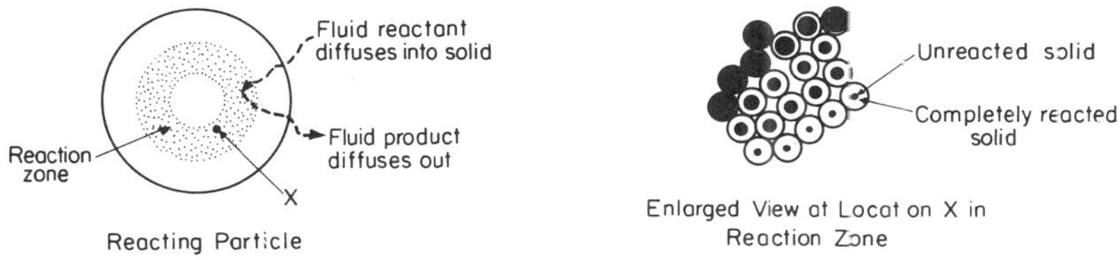

FIG. 5.4. Grain model for the reaction of a porous solid.[1] Reprinted with permission from *IEC Proc. Des. Dev.*, **19**, 118 (1980) © 1980 American Chemical Society.

appeared as a *boundary condition* on the diffusion equation.) The local rate of reaction within the solid is related to the fluid reactant concentration in the interstices between the grains by considering that each grain reacts in a shrinking core fashion. The differential equation for the diffusion of fluid reactant requires two boundary conditions. The one at the external surface of the lump (pellet) is the same as in the shrinking core model. The other condition arises from symmetry and is that the concentration gradient is zero at the center of the lump. Stoichiometry yields the rate at which the reaction front radius within each grain is shrinking as a function of the local concentration of fluid reactant. Generally the grain model equations require a numerical solution, although exact analytical solutions are possible in some circumstances and an approximate analytical solution exists relating the extent of reaction to time.[2]

It should be noted that before either the shrinking core model or the grain model can be incorporated into a model for a fluid–solid reactor a great deal of experimental work is usually necessary. The reaction rate constant for a fluid–solid reaction is not a well-defined physical quantity in the sense of, say, the thermal conductivity of steel or the viscosity of water. The rate constant, besides being particularly sensitive to temperature, is frequently dependent on trace impurities in the solid and sometimes on the mechanical history of the solid. Thus, for example, it would be dangerous to use, say, a rate constant for iron ore reduction determined for samples from one ore body to predict the behavior of a reactor in which a different ore is reduced, even though the more sophisticated models permit a correct allowance for variations in solid microstructure (porosity variations, etc.). The mathematical models for the reaction of a single entity of solid play an important role in such experimental work, since it is usual (and sensible) to carry out experiments on single lumps or pellets of solid. The models can then be used to obtain the reaction rate constant and effective diffusivities for the solid under study by fitting the model to the experimental data.

5.2. MODELS FOR FLUID–SOLID REACTORS

For industrial reactors the concentration of the fluid reactant is usually a function of position in the reactor and of time. The models for the reaction of individual entities of solid cannot therefore be applied directly, since they are developed for a constant environment surrounding the solid. An additional complication arises if the temperature of the solid varies with position or time.

For reactors it is therefore necessary to employ additional equations describing the variation of fluid reactant concentration (and, for the nonisothermal case, of temperature) within the reactor. Such equations are transport equations for a particular fluid species (and for heat) and will contain convective terms arising from the transport of the species (or heat) by the bulk motion of the fluid. A preliminary step is therefore to arrive at some quantitative description of the flow of the fluid within the reactor.

5.2.1. Isothermal Packed Beds

At this stage it is worthwhile to examine a particular example of a reactor in order to further illustrate these points. Let us consider a packed bed reactor in which a solid is reacting with a gas. The reactor in the HyL process is an example of such a reactor. In Section 5.2.4 we will examine the question of fluid flow in packed beds, particularly with regard to flow maldistribution.

Let us, for the moment, use a simpler approach and assume that the gas moves through the bed in plug flow. In addition it will be assumed that the reaction of an individual solid pellet can be represented by the grain model, that the solid pellets are all of the same size, and that the bed is isothermal. (Mention will be made later of models in which the last two assumptions are avoided).

As has been mentioned already, within the packed bed reactor a particular pellet will be immersed in a gas stream whose concentration will vary with time and the grain model is not directly applicable. For first-order reactions it is, however, possible to write:[3]

$$\frac{dX}{dt^*} = \theta\beta \tag{15}$$

X is the extent of reaction of the particular pellet (and of all other pellets at the same distance from the gas inlet) and t^* is time, rendered dimensionless by the following definition:

$$t^* = \left(\frac{bk\bar{C}_{A0}A_g}{\rho_s F_g V_g}\right) t \tag{16}$$

where b is the stoichiometric coefficient (moles of solid reacting per mole of gaseous reactant reacting),

k is the first-order rate constant,

A_g, F_g, and V_g are the area, shape factor, and volume, respectively, of a grain, and

ρ_s is the density of the solid reactant.

TABLE 5.1. β Values for Calculations on Reaction of a Packed Bed of Solid with a Gas

Extent of Reaction	$F_g = 3$						$F_g = 2$						$F_g = 1$					
X	$\hat{\sigma} = 0.3$	0.5	0.8	1.2	1.8	3.0	0.3	0.5	0.8	1.2	1.8	3.0	0.3	0.5	0.8	1.2	1.8	3.0
0.00	2.72	2.37	1.88	1.42	1.02	0.653	1.87	1.69	1.40	1.10	0.808	0.522	0.966	0.912	0.812	0.679	0.526	0.353
0.05	2.63	2.30	1.82	1.38	0.984	0.614	1.82	1.65	1.37	1.07	0.787	0.503	0.966	0.912	0.812	0.679	0.526	0.353
0.10	2.55	2.23	1.77	1.33	0.942	0.575	1.78	1.61	1.34	1.05	0.764	0.481	0.966	0.912	0.812	0.679	0.526	0.353
0.15	2.46	2.16	1.71	1.28	0.898	0.533	1.73	1.57	1.31	1.02	0.740	0.458	0.966	0.912	0.812	0.679	0.526	0.353
0.20	2.37	2.08	1.65	1.23	0.853	0.490	1.68	1.53	1.28	0.993	0.715	0.434	0.966	0.912	0.812	0.679	0.526	0.353
0.25	2.28	2.01	1.59	1.18	0.806	0.444	1.63	1.49	1.24	0.964	0.689	0.406	0.966	0.912	0.812	0.679	0.526	0.353
0.30	2.18	1.93	1.53	1.13	0.758	0.396	1.58	1.44	1.20	0.933	0.660	0.375	0.966	0.912	0.812	0.679	0.526	0.353
0.35	2.08	1.85	1.47	1.07	0.707	0.347	1.52	1.39	1.17	0.901	0.630	0.339	0.966	0.912	0.812	0.679	0.526	0.353
0.40	1.98	1.76	1.40	1.02	0.655	0.297	1.47	1.35	1.13	0.867	0.596	0.297	0.966	0.912	0.812	0.679	0.526	0.298
0.45	1.88	1.68	1.33	0.959	0.600	0.251	1.41	1.29	1.08	0.831	0.560	0.251	0.966	0.912	0.812	0.679	0.526	0.243
0.50	1.77	1.58	1.26	0.898	0.542	0.213	1.35	1.24	1.04	0.791	0.519	0.212	0.966	0.912	0.812	0.679	0.526	0.228
0.55	1.65	1.49	1.19	0.833	0.483	0.182	1.28	1.18	0.992	0.749	0.474	0.181	0.966	0.912	0.812	0.679	0.515	0.187
0.60	1.54	1.39	1.11	0.765	0.422	0.155	1.21	1.12	0.941	0.703	0.420	0.155	0.966	0.912	0.812	0.679	0.414	0.156
0.65	1.41	1.28	1.02	0.693	0.361	0.132	1.13	1.05	0.885	0.651	0.359	0.132	0.966	0.912	0.812	0.679	0.352	0.133
0.70	1.28	1.17	0.930	0.617	0.306	0.112	1.05	0.979	0.823	0.592	0.304	0.112	0.966	0.912	0.812	0.654	0.289	0.116
0.75	1.14	1.04	0.832	0.535	0.257	0.0941	0.964	0.899	0.754	0.524	0.255	0.0937	0.966	0.912	0.812	0.530	0.247	0.0949
0.80	0.986	0.909	0.724	0.448	0.212	0.0779	0.865	0.809	0.675	0.440	0.210	0.0779	0.966	0.912	0.812	0.453	0.216	0.0775
0.85	0.819	0.760	0.603	0.357	0.170	0.0626	0.752	0.705	0.581	0.348	0.168	0.0625	0.966	0.912	0.672	0.354	0.167	0.0612
0.90	0.629	0.587	0.461	0.265	0.129	0.0477	0.616	0.578	0.460	0.259	0.127	0.0477	0.966	0.912	0.504	0.264	0.129	0.0470
0.95	0.399	0.374	0.284	0.166	0.0843	0.0319	0.437	0.406	0.285	0.162	0.0828	0.0319	0.966	0.658	0.319	0.168	0.0834	0.0317
0.99	0.137	0.124	0.0877	0.0581	0.0340	0.0144	0.193	0.151	0.0943	0.0588	0.0334	0.0142	0.488	0.243	0.118	0.0689	0.0357	0.0151

$\theta = C_{Ao}/\bar{C}_{Ao}$ where C_{Ao} is the concentration of the gaseous reactant external to the pellet and $\bar{C}_{Ao}$ is the concentration at the gas inlet of the bed. β is an effectiveness factor for the gas–solid reaction and is given in Table 5.1, assuming mass transfer external to the pellet presents negligible resistance to the progress of reaction. The β factor may also be obtained by differentiating the Evans-Ranade approximate solution to the grain model equations referred to in Section 5.1. The β factor is a function of both the extent of reaction and a reaction modulus $\hat{\sigma}$, which is

$$\hat{\sigma} = \frac{V_p}{A_p}\sqrt{\frac{F_p(1-\varepsilon_p)k}{2D_e}\frac{A_g}{V_g}} \tag{17}$$

where V_p is the volume of a pellet,
A_p is its external surface area,
F_p is its shape factor,
ε_p is its porosity, and
D_e is the effective diffusivity of the gaseous reactant within the pores of the solid.

Equation (15) gives the rate of reaction at some position in the bed as a function of the concentration of gaseous reactant in the interstices between pellets at that point, C_{AO} (or in dimensionless form, θ). It remains to obtain an equation describing the concentration in the interstices as a function of position. Such an equation is obtained by carrying out a mass balance on a differential slice of the packed bed. The result is

$$-G_g\frac{\partial\theta}{\partial y} = \frac{(1-\varepsilon_p)(1-\varepsilon_v)\rho_s}{b}\frac{dX}{dt} + (\varepsilon_p + \varepsilon_v)\bar{C}_{Ao}\frac{\partial\theta}{\partial t} \tag{18}$$

where G_g is the superficial mole velocity of the gaseous reactant (superficial linear velocity of the gas stream multiplied by the moles per unit volume of gaseous reactant at the gas inlet),
y is the distance along the bed measured from the gas inlet, and
ε_v is the void fraction of the bed.

The equation assumes that diffusive (dispersive) transport of the gaseous reactant in the direction of gas flow is negligible compared to convective transport. The last term on the right of this equation represents the accumulation of gaseous reactant in the interstices and can safely be neglected, compared to the other terms. Switching to dimensionless variables and employing Eq. (15), we get

$$-\frac{\partial\theta}{\partial y^*} = \frac{\partial X}{\partial t^*} = \theta\beta \tag{19}$$

where y^* is a dimensionless distance along the bed defined by

$$y^* = \frac{kA_g\bar{C}_{Ao}}{G_gF_gV_g}(1 - \varepsilon_p)(1 - \varepsilon_v)y \tag{20}$$

Equation (19) can readily be manipulated to give

$$\frac{\partial X}{\partial t^*} = \beta \exp\left[-\int_0^{y^*} \beta \, d\lambda\right] \tag{21}$$

where λ is a dummy variable. This equation makes sense in that the rate of reaction at some point y^* in the bed is seen to be a function of the extent of reaction at that point (through the dependence of β on X) and also on the reaction of solid upstream ($0 \leqq \lambda \leqq y^*$) from y^*, such reaction upstream affecting the concentration of gaseous reactant at y^*. Equation (21) can be solved numerically to give X as a function of y^* and t^*; the results appear in Figures 5.5–5.7. An alternative presentation of the results is possible, as shown in Figure 5.8. The extent of reaction in the bed is plotted as a function of distance

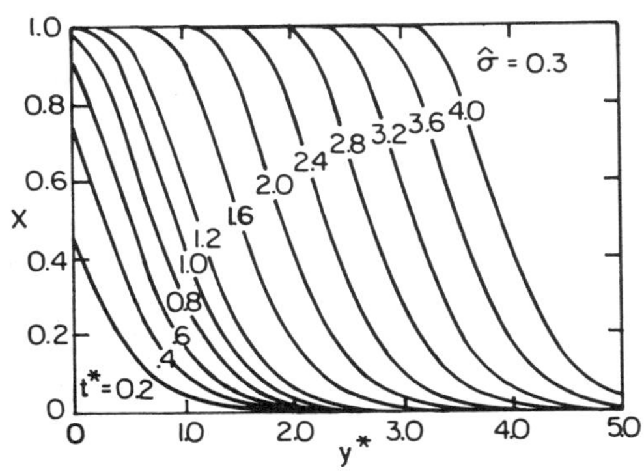

FIG. 5.5. Fixed bed reactors. Extent of reaction versus position in bed for reaction modulus = 0 3.[3] Reprinted with permission from *IEC Proc. Des. Dev.*, **13**, 149 (1974) © 1974 American Chemical Society.

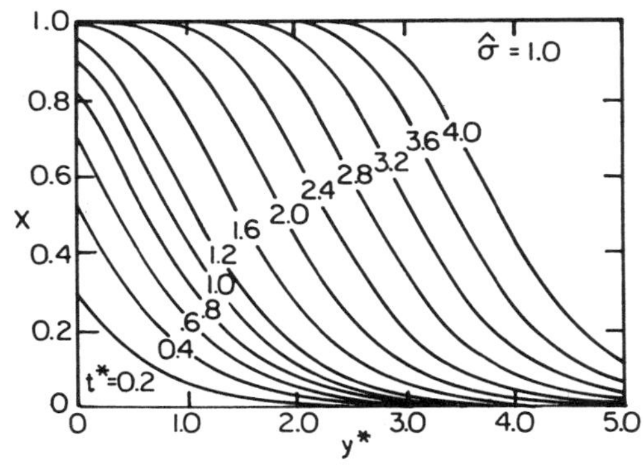

FIG. 5.6. Fixed bed reactor. Extent of reaction versus position in bed for reaction modulus = 1.0.[3] Reprinted with permission from *IEC Proc. Des. Dev.*, **13**, 149 (1974) © 1974 American Chemical Society.

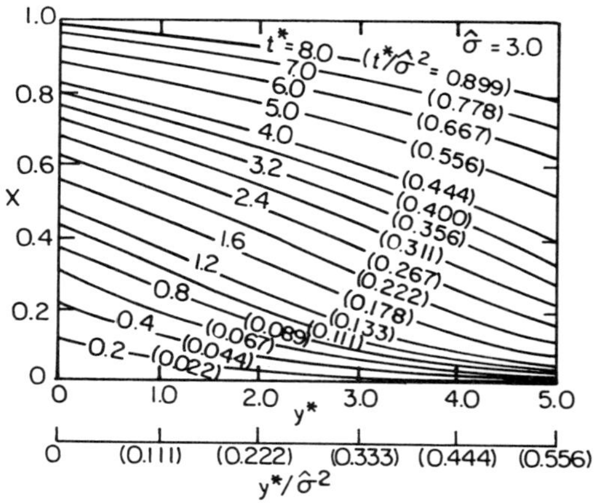

FIG. 5.7. Fixed bed reactor. Extent of reaction versus position in bed for reaction modulus = 3.0.[3] Reprinted with permission from *IEC Proc. Des. Dev.*, **13**, 149 (1974) © 1974 American Chemical Society.

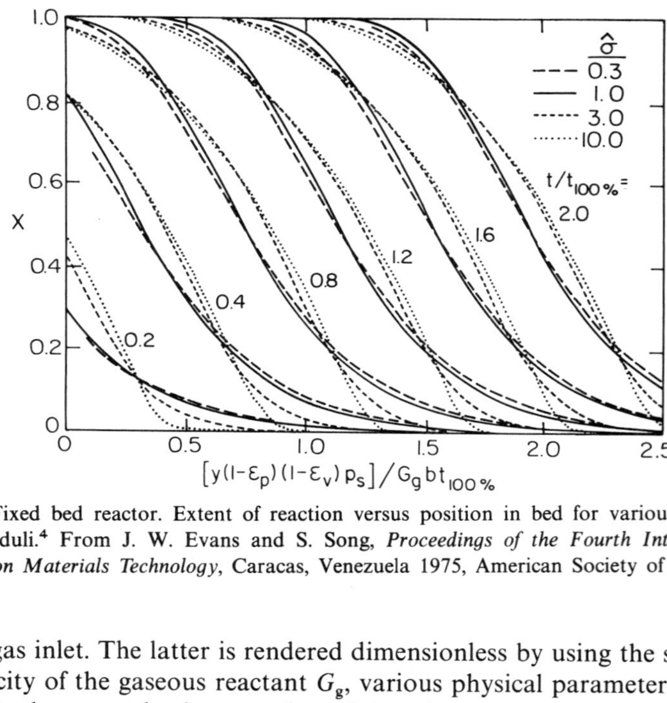

FIG. 5.8. Fixed bed reactor. Extent of reaction versus position in bed for various times and reaction moduli.[4] From J. W. Evans and S. Song, *Proceedings of the Fourth Inter-American Conference on Materials Technology*, Caracas, Venezuela 1975, American Society of Mechanical Engineers.

from the gas inlet. The latter is rendered dimensionless by using the superficial mass velocity of the gaseous reactant G_g, various physical parameters, and the time required to completely react the solid at the gas inlet conditions $t_{100\%}$, a quantity that can be determined easily by experiments on single pellets. It is seen that after an initial period a reaction zone moves down the bed with an unchanging profile and a velocity of

$$\frac{G_g b}{\rho_s(1-\varepsilon_p)(1-\varepsilon_v)} \tag{23}$$

which is to be expected from stoichiometry. Furthermore, it should be noted that, as represented in the manner of Figure 5.8, the results are only slightly dependent on the value of the reaction modulus, $\hat{\sigma}$.

EXAMPLE 2

Calculate the extent of reaction at the exit of an isothermal packed bed reactor, in which iron oxide pellets are being reduced to iron, 10 min after start-up of the reactor. The reaction can be assumed to be

$$Fe_3O_4 + 4H_2 = 3Fe + 4H_2O$$

and to be proceeding under pore diffusion control. The porosity of the pellets is 25%, and the void fraction of the bed is 40%. The bed length is 20 cm, and the superficial molar velocity of the gas is 12 mole/m^{2}s. Laboratory experiments have determined that the time to completely react a sphere at entrance gas conditions is 300 s. The density of Fe_3O_4 is 5.18 g/cm^3, and its molecular weight is 231.5.

Solution.

The molar density of nonporous Fe_3O_4, $\rho_s = 2.24 \times 10^4$ moles/m^3. For the gas exit ($y = 0.2$ m), the abscissa in Figure 5.8 becomes

$$\frac{0.2 \text{ m}(0.75)(0.6)2.24 \times 10^4 \text{ moles/m}^3}{12 \text{ mole/m}^2\text{s} \times 0.25 \times 300} = 2.24$$

For pore diffusion control $\hat{\sigma}$ is large, and we can use the curve for $\hat{\sigma} = 10$ in Figure 5.8. Since $t/t_{100\%} = 2$, we read $X = 0.33$ at the bed exit from this curve. Note from the curve that approximately half of the bed has completely reacted at this time.

It is worthwhile reviewing the features of the model discussed here. The model describes the reaction of an individual pellet by using the grain model and assumes plug flow of the gas in the bed, that the bed is isothermal, and that all pellets are the same size. The model is seen to be a little simplistic, and development of more sophisticated models is called for. A first step in this direction is discussed in Section 5.2.2.

5.2.2. Nonisothermal Packed Beds

It is a straightforward matter to allow the temperature of the reactor to be a function of position in the bed (while constant in time). This improvement in the model is described by Ranade and Evans.[5] The improved model also makes allowance for the reversibility of the reaction and for total pressure variations in the bed. The investigation described by Ranade and Evans involved experimental work on the reduction of taconite pellets by hydrogen. The results of the model predictions for the behavior of a nonisothermal bed of taconite pellets undergoing reduction with hydrogen were compared with the experimental

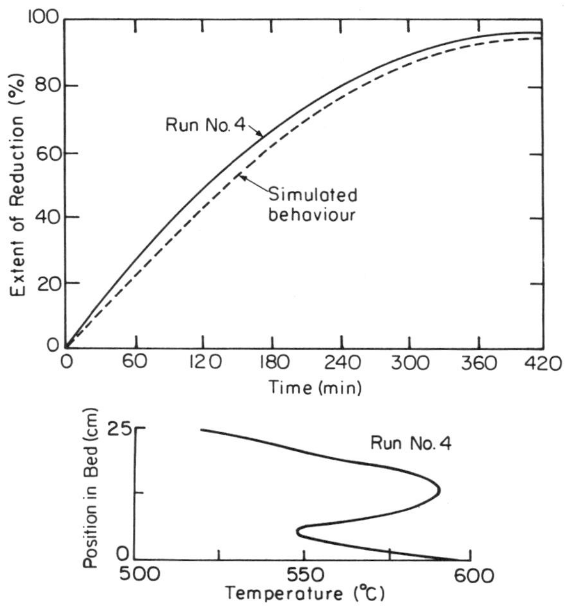

FIG. 5.9. Temperature profile and overall extent of reaction for reduction of iron ore pellets in a nonisothermal packed bed reactor.[5] Reprinted with permission from *IEC Proc. Des. Dev.*, **19**, 121 (1980) © 1980 American Chemical Society.

measurements (Figure 5.9). Kinetic parameters for use in the simulations were obtained in a separate series of experiments in which individual taconite pellets were reduced in a gravimetric apparatus, such as described in Chapter 4. It is seen that even though the bed had temperature variations of up to approximately 100°C, the model was able to predict the overall conversion of the bed as a function of time.

5.2.3. Packed Beds with a Distribution of Solid Sizes

So far we have considered the reaction of packed beds where the solid entities (pellets) making up the bed all have roughly the same size. This is a reasonable approximation for the reaction of pellets of, say, iron ore, which are manufactured in a fairly close size range in mineral processing operations. In some instances, however, a broader size range might be expected, for example, in the column leaching of lumps of a metal ore, where a reactor might be expected to contain large lumps along with fine particles. Such a reactor will now be treated assuming isothermal behavior and using the shrinking core model to describe the reaction of an individual lump of ore.[6]

The rate of reaction of a lump of ore under such circumstances is given by

$$\frac{dX_i}{dt} = \frac{bkC_{\mathrm{Ao}}}{\rho_s r_i}\left(\frac{3f_i}{1 + \sigma_i^2 f_i - 6\sigma_i^2 f_i^2}\right) \tag{24}$$

where X_i is the extent of reaction of lump of radius r_i, and
σ_s is the number of moles of leachable solid per unit volume of one lump.

$$f_i = (1 - X_i)^{1/3} \tag{25}$$

The reaction modulus of size i is σ_i, given by

$$\sigma_i^2 = \frac{kr_i}{6D} \tag{26}$$

where D is the effective diffusivity of the lixiviant in the ash layer and other symbols are as before.

Introducing a dimensionless time,

$$t^* = \frac{bk\bar{C}_{Ao}t}{\rho_s r_0} \tag{27}$$

Equation (24) becomes

$$\frac{dX_i}{dt^*} = \theta\beta_i \tag{28}$$

where

$$\beta_i = \frac{3f_i^2\sigma_0^2}{(1 + 6\sigma_i^2 f_i - 6\sigma_i^2 f_i^2)\sigma_i^2} \tag{29}$$

is an effectiveness factor for size i and r_0 is a reference size of corresponding modulus σ_0. A mass balance for the lixiviant over a differential slice of the column gives

$$\frac{-\dot{V}r_0}{A(1-\varepsilon_v)k\bar{C}_{Ao}}\frac{\partial C_{Ao}}{\partial y} = \frac{\rho_s r_0}{bk\bar{C}_{Ao}}\sum_i w_i \frac{dX_i}{dt} \tag{30}$$

where $\dot{V}$ is the volumetric flow rate of the fluid,
A is the cross-sectional area of the column, and
w_i is the weight fraction of lumps of size i.

Letting

$$y^* = \frac{A(1-\varepsilon_v)ky}{\dot{V}r_0} \tag{31}$$

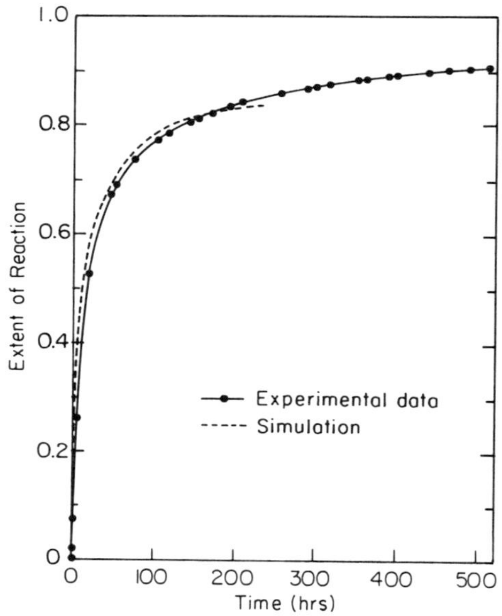

FIG. 5.10. Overall extent of leaching versus time for the sulfuric acid leaching of a column of copper oxide ore.[6]

and manipulating the above equations yields

$$\frac{dX_i}{dt^*} = \beta_i \exp\left\{-\int_0^{y^*} \sum_i w_i \beta_i \, d\lambda\right\} \tag{32}$$

Equation (32) should be compared to Eq. (21), the single-size case. The difference between the two equations is the summation appearing in Eq. (32), which arises because the fluid reactant concentration at y^* is diminished by the reaction of *all* lump sizes upstream from y^*.

Equation (32) can be solved numerically. The solution is obtained at a discrete set of y^* values along the bed and for each of several lump sizes at each position. For each size at each position, Eq. (32) is integrated numerically (e.g., by the Runge–Kutta technique), the integral on the right of Eq. (32) being obtained numerically (e.g., by the trapezoidal rule). The overall extent of reaction of the column is then given by:

$$X = \int_0^{L^*} \left(\sum w_i X_i\right) dy^* \tag{33}$$

where L^* is the dimensionless column length.

In Figure 5.10, the results of this simulation are compared with experiments in which a column of copper oxide ore has been leached with dilute sulfuric acid. Kinetic parameters for use in the simulation were obtained by beaker experiments in which a few lumps of ore were leached in an excess of acid.

5.2.4. Reactors with Spacial Variation of Gas and Solids Flow

For another example of a more detailed model of a fluid–solid reactor, we shall consider the work of Yagi and Szekely,[7,8] who have constructed a mathematical model of a countercurrent moving bed gas–solid reactor wherein iron ore is reduced by hydrogen. Such a model would apply to a continuous, shaft-type, direct reduction unit and also provides some insight into the behavior of the shaft region of a blast furnace where, however, reduction is by carbon monoxide. The principal objective of the model was the examination of the effects of nonuniform gas and solids flow on the overall behavior of a countercurrent moving bed reactor.

The model used for the reaction of an individual lump of ore in this case is the three-interface model of Hara et al.[9] This model bears some similarity to the shrinking core model except that there are three moving interfaces within the lump, at which the three reactions

$$3Fe_2O_3 + H_2 = 2Fe_3O_4 + H_2O \tag{34}$$

$$Fe_3O_4 + H_2 = 3FeO + H_2O \tag{35}$$

and

$$FeO + H_2 = Fe + H_2O \tag{36}$$

take place.

Yagi and Szekely developed their model for a cylindrical bed operating at steady state. Temperature within a given solid lump was assumed to be uniform. The governing differential equations are too complex for reproduction here but they consist of

1. A material balance on the hydrogen gas allowing for convection of hydrogen in both axial and radial directions together with radial dispersion. The equation takes into account the expansion of the gas by both pressure and temperature gradients in the bed as well as for the consumption of hydrogen by reaction.
2. A material balance on the solid lumps that was simplified by assuming that the solids move only in the axial direction (though with a velocity that would depend on radial position).
3. A heat balance on the gas allowing for convection in both axial and radial directions, together with radial dispersion.
4. A heat balance on the solid that allows for axial and radial conduction of heat, together with axial convection by the moving solids, heat transfer to the gas stream, and heat generated within the solid by the gas.
5. The Ergun equation, written in two dimensions to describe the flow of gas through the bed under a pressure gradient.

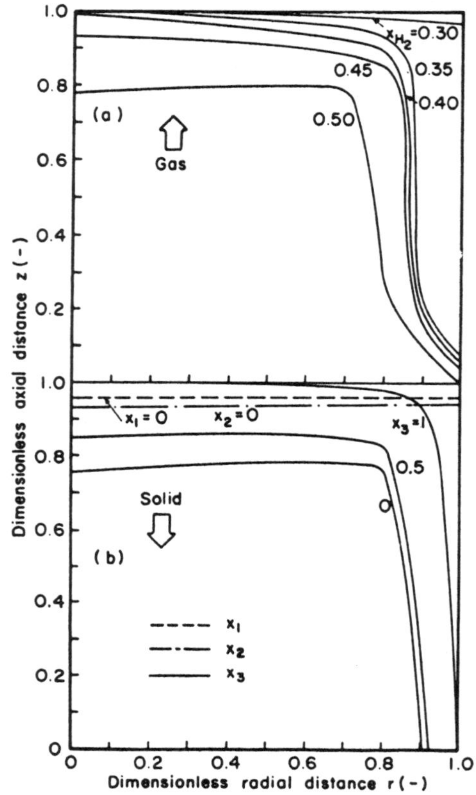

FIG. 5.11. Contour lines for various extents of reaction cpmputed by Yagi and Szekely for a countercurrent moving bed reactor in which iron ore is reduced. Calculations for nonuniform distribution of gas.[7]

Physical and chemical constants used in the model were obtained from the literature, and these were used in conjunction with the differential equations and boundary conditions in a numerical solution. The complexity of the model is evident from the computer time required for solution: 900–2000 s on a CDC 4000.

Typical results of the computations appear in Figures 5.11 and 5.12. The upper half of each figure shows the lines of constant mole fraction of hydrogen, while the lower half of each figure provides information on the amount by which the solid has reacted. $X_1 = 0$ indicates the completion of the reduction of hematite to magnetite; $X_3 = 1$ corresponds to the start of the reduction of wustite to iron; and $X_3 = 0$ corresponds to its completion. The case under consideration is one where the void fraction of the bed and its particle size decrease outward from the center, implying a greater resistance to gas flow closer to the wall. The linear velocity of the solids is set constant. Figure 5.12 is a recomputation of Figure 5.11 with the gas flow assumed to be uniform (i.e., only

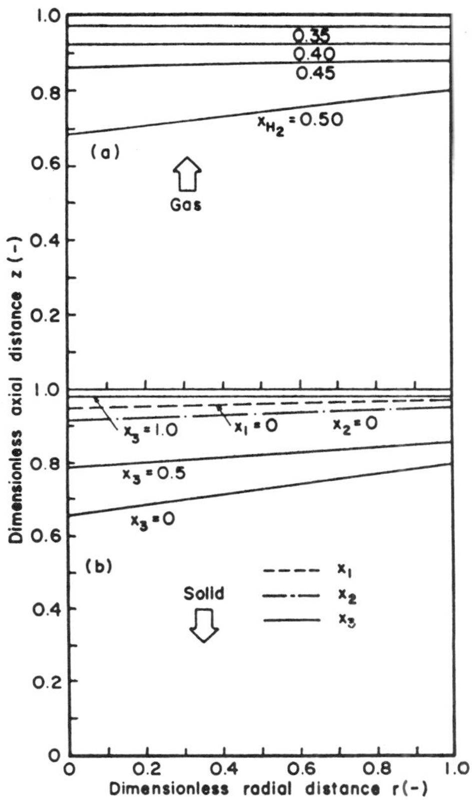

FIG. 5.12. As in Figure 4.11 but with uniform gas distribution.[7]

in the axial direction). This assumption had been made by many previous investigators and is clearly seen to be unsatisfactory.

The shapes of the concentration isopleths in Figure 5.11 are as might be expected. The greater resistance to gas flow close to the wall results in substantially reduced gas flow throughout that region and, consequently, reduced reaction rate. Indeed, it is seen that the solid is only partly reduced in this region ($X_3 > 0$) when it leaves the bed. It should be noted that calculations based on the assumption of plug flow of the gas (Figure 5.11) do *not* reveal the incomplete reaction of the solid.

Quite different behavior is observed in the case of a bed where the porosity and solid size *increase* toward the wall. In this case (see Figure 5.13) the center of the bed has a high resistance to gas flow, and consequently the reaction is starved of gas at the axis of the bed. Again, incompletely reacted solid leaves the bed. In this case also the recomputation of the results assuming (incorrectly) a purely axial flow of gas gave quite different isopleths and failed to reveal that part of the solids discharging from the bed would be incompletely reacted.

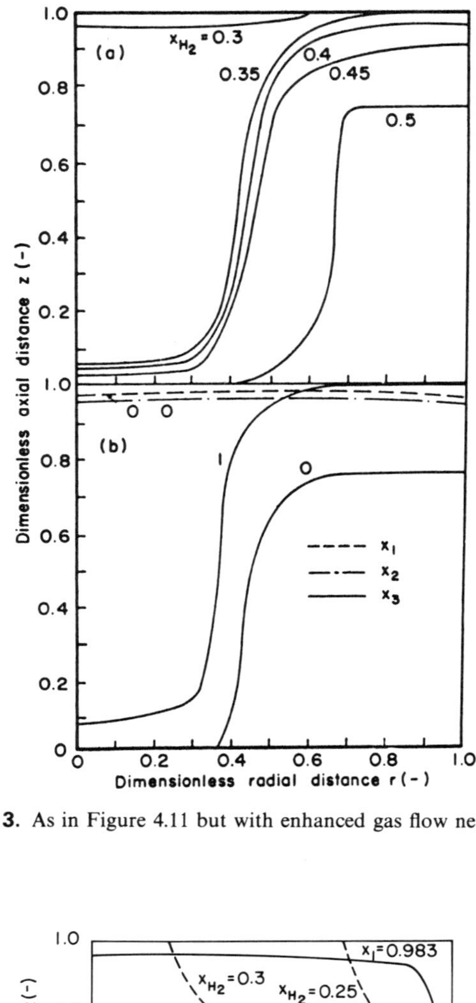

FIG. 5.13. As in Figure 4.11 but with enhanced gas flow near the walls.[7]

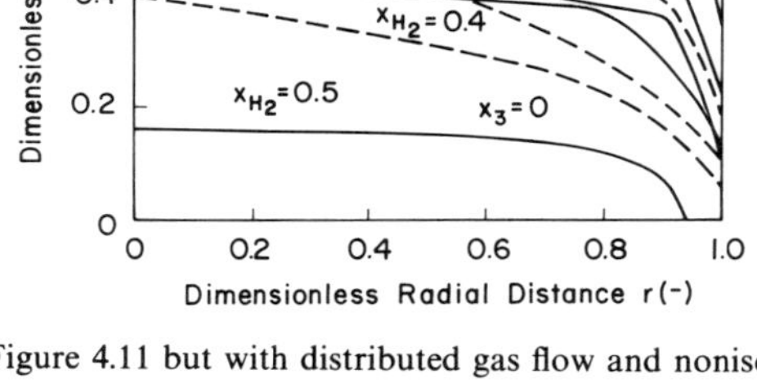

FIG. 5.14. As in Figure 4.11 but with distributed gas flow and nonisothermal conditions.[8]

The results presented were for isothermal conditions. Yagi and Szekely published some results for nonisothermal conditions, for example, those shown in Figure 5.14, which pertains to a bed where the gas flow resistance increases toward the wall.

5.2.5. Models for Rotary Kilns

Another example of a mathematical model of a gas–solid reactor can be found in the work of Venkateswaran and Brimacombe on the SL/RN direct reduction process.[10] The reactor shown in Figure 5.15 consists of a rotary kiln into which coal (and recycle char) and iron ore are fed. In the solids bed, the coal reacts with the iron ore in the solid state, via the Boudouard reaction:

$$C + CO_2 = 2CO \tag{37}$$

$$Fe_2O_3 + 3CO = 2Fe + 3O_2 \tag{38}$$

Some of the CO gas resulting from Eq. (37) transfers from the bed to the freeboard space, where it is combusted with air that is injected axially into the kiln through several ports. The combustion provides the process with heat to raise the solids to the reaction temperature of 900–1000°C and to drive the Boudouard reaction, which is strongly endothermic. Thus, the SL/RN process consists of two completely different reaction zones: the bed, which is highly reducing, and the freeboard, which is very oxidizing.

In a process such as the SL/RN, one is interested in predicting the temperature and concentration distribution in the two zones for several reasons.

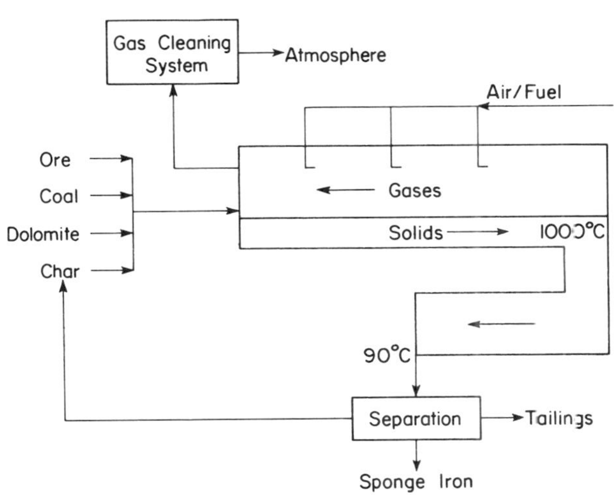

FIG. 5.15. Schematic diagram of the SL/RN process. From *Met. Trans.* **8B**, 387 (1977).

First, one would like to know the length of kiln required to achieve a desired percent metallization of the product iron for various coals and ores that have dissimilar kinetic behavior. Second, a major control parameter is the distribution of air injected through the ports over the length of the kiln; and one would like to know what the optimum distribution should be to maximize kiln productivity and to minimize such problems as local overheating and accretion buildup on the walls. Third, it would be desirable to determine the effect of coal reactivity or ore reducibility on kiln operation.

From the foregoing description, mass transfer, chemical kinetics, and heat transfer must be included in the mathematical model. Momentum flow is also important, but reasonable assumptions can be made about mixing so that the momentum equation can be ignored. Thus, axial plug flow is assumed individually for the solids bed and freeboard gases, which are moving countercurrently. In the radial direction, the bed and freeboard gas are each taken to be backmixed such that there are no radial gradients of temperature or concentration in the two phases. The latter assumption may not hold well for deep beds but, in the absence of either empirical data or valid models of bed mixing, is necessary. Even in the case of shallow beds, radial gradients may exist if the bed slips against the wall rather than continuously turning over as the kiln rotates. Henein et al.[11] have delineated the conditions under which slipping, and other forms of bed motion in rotary kilns, dominate.

Under the assumed mixing conditions, temperature and concentration gradients exist only in the axial direction. Thus, the mathematical formulation of the model is based on heat and material (molar) balances over an infinitesmal section element that includes the solids, freeboard gas, and kiln wall. The molar balances must be made on each of the important chemical species (e.g., carbons and Fe_2O_3 in the bed and CO, CO_2, H_2, N_2, and O_2) in the freeboard. However, in making heat balances, it may be assumed that all species in the solid phase are at the same local temperature; and obviously this is so in the freeboard. Therefore, only two equations result. The rates of chemical reaction (e.g., reduction, Boudouard reaction, coal pyrolysis, and combustion) are included in the heat and material balances and thus must be characterized quantitatively. Also, heat exchange by radiation and convection among the exposed kiln wall, freeboard gas (including flames), and bed must be formulated mathematically. This subject has been taken up recently by Gorog et al.[12] The set of first-order ordinary differential equations that results from the balances represents a boundary value problem because the inlet conditions of the solids are specified at one end of the kiln, while the inlet conditions of the counterflowing gases are specified at the other. Clearly, a numerical solution is necessary.

A set of model predictions of axial temperature and concentration (or partial pressure) gradients for a typical set of operating conditions in a 100-tpd pilot plant is shown in Figures 5.16 and 5.17 respectively. Thus, the solids temperature is seen to rise from the inlet value and to reach a plateau that corresponds to the reduction zone. The relatively constant solids temperature through this region is due, of course, to the endothermic Boudouard reaction. The wall

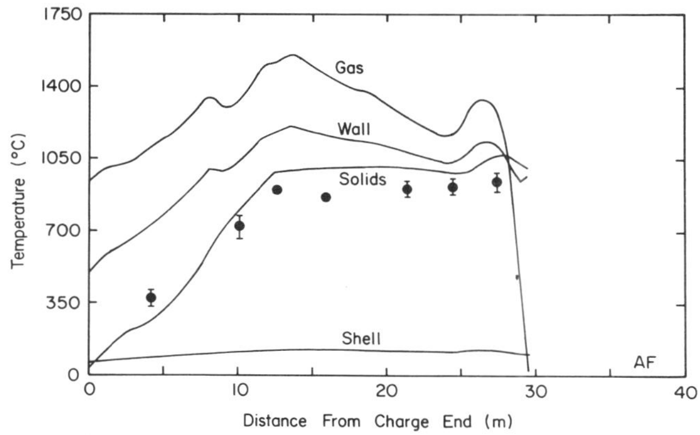

FIG. 5.16. The temperature profiles for gas, wall, solids, and shell calculated by Gorog et al. The dots are measured solids temperatures in a pilot kiln.[12]

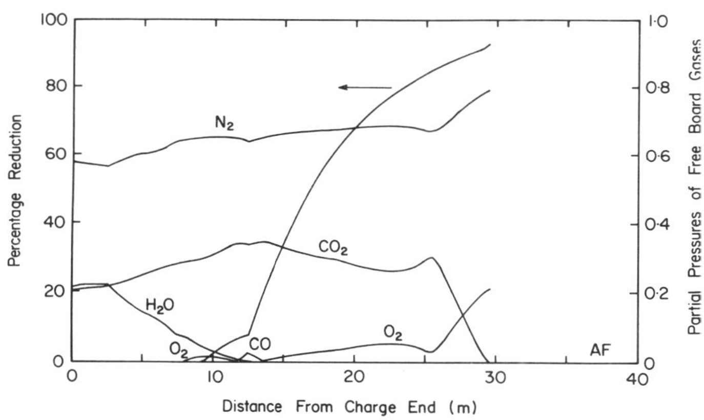

FIG. 5.17. Composition and percentage reduction profiles for a pilot kiln from the model of Gorog et al.[12]

temperature profile is also predicted by the model and exhibits a peak close to the beginning of the reduction zone, where the freeboard gas temperature is similarly a maximum. The high temperature in this central part of the kiln is caused by the initial rapid rates of reduction and coupled Boudouard reaction with a copious evolution of CO from the bed. Accretion formation is most likely in this zone, owing to the elevated temperatures, and this is borne out in practice.

In Figure 5.17 an excess of oxygen in the freedboard is predicted through the reduction zone, which suggests that the flames, due to burning CO, should be close to the bed surface (assuming combustion is virtually instantaneous), which should enhance heat transfer to the bed. This prediction also agrees with observation of the bed surface. In the preheat zone, where the solids temperatures are rising, the partial pressures of oxygen and combustible gases (CO and volatiles from the coal) are predicted to be negligible, which indicates stoichiometric combustion in the freeboard. Of course, under these conditions the gas temperature and heat transfer to the bed are maximized.

The model predictions further revealed the sensitivity of the SL/RN process to the air profile. Too much or too little air in the reduction zone has a very deleterious effect on bed temperature and kiln productivity. Moreover, the model showed that the best coals to be used as reductant are highly reactive (e.g., lignite and subbituminous coals). The high reactivity results in lower bed temperature, fewer accretion problems, and higher productivity.

5.3. CONCLUDING REMARKS

The modeling of fluid–solid reactors represents a challenging task for the modeler of metallurgical processes. The amount of data available on reaction rates is small (in contrast, say, to data on electrical conductivities or vapor pressures, for which extensive tables and reliable correlations can be found). Consequently, laboratory experiments are frequently an essential adjunct to model development. Such experimentation can be guided by the models for heterogeneous reactions described in this chapter. Additional complications involved in reactor modeling are due to variations in solids characteristics (composition, porosity, particle size, etc.), which are inherent in processing materials obtained from nature; the nonisothermal behavior encountered in many of these reactors; complications due to maldistribution of the flow of the fluid reactant; and complications due to peculiarities of the reactor. An example of the last named was mentioned in this chapter: the mixing (or segregation) of solids that occurs in rotary kilns. The modeling of fluid–solid reactors has been an active area of research in the past decade and, because of its complexity, is likely to remain so for years to come.

REFERENCES

1. J. Szekely, J. W. Evans, and H. Y. Sohn, *Gas–Solid Reactions*, Academic Press, New York, 1976, chap. 3.
2. J. W. Evans and M. Ranade, *Chem. Eng. Sci.*, **35**, 1261 (1980).
3. J. W. Evans and S. Song, *IEC Proc. Des. Dev.*, **13**, 146 (1974).
4. J. W. Evans and S. Song, Proceedings of the Fourth Inter-American Conference on Materials Technology, ASME, Caracas, Venezuela, 1975.
5. J. W. Evans and M. Ranade, *IEC Proc. Des. Dev.*, **19**, 118 (1980).
6. J. F. Livingston, Master's thesis, University of California, Berkeley, 1979.
7. J. Yagi and J. Szekely, *Trans. ISI Japan*, **17**, 569 (1977).

8. J. Yagi and J. Szekely, *AIChE J.*, **25**, 800 (1979).
9. Y. Hara, M. Tsuchiya, S. Kondo, et al., *Tetsu-to-Hagane*, **60**, 1261 (1974).
10. V. Venkateswaran and J. K. Brimacombe, *Met. Trans.*, **8B**, 397 (1977).
11. H. Henein, J. K. Brimacombe, and A. P. Watkinson, *Met. Trans.*, **14B**, 207 (1983).
12. J. P. Gorog, J. K. Brimacombe, and T. N. Adams, *Met. Trans.*, **12B**, 55 (1981); **13B**, 153 (1982); **14B**, 411 (1983).

6 LADLE METALLURGY OPERATIONS

During the past decade there has been a growing interest in the development and application of ladle metallurgy for the finishing treatment of molten steel. Indeed, some form of ladle metallurgy step, such as argon stirring in order to achieve homogenization, has become an established component of the overall steelmaking sequence.[1-4] A sketch of typical ladle metallurgy operations is shown in Figure 6.1, which depicts in a historical sequence of the evolution of argon-stirred ladles, vacuum degassing systems, and induction furnaces.

In general the operation of these systems involves the agitation of melts that may range from 5 to 250 tons in size, using injected-gas streams, vacuum-driven circulation systems, and electromagnetic force fields. In addition to agitation, which brings about homogenization, chemical and physical changes are also often effected, such as

desulfurization, by injected lime, CaC_2, or, magnesium;
deoxidaton, by Si or Al addition;
degassing, by using vacuum or purge gases; and
coalescence of inclusion particles.

Ladle metallurgy operations constitute excellent illustrations of the modeling approach because in this field mathematical modeling, physical modeling, and plant-scale measurements have been successfully combined to allow the rapid commercialization of a new technology. The problems encountered in ladle metallurgy are also helpful in illustrating how the specific nature of a problem and an investigator's expertise may determine the optimal approach to selecting mathematical modeling, physical modeling, and plant (pilot-scale) measurements as complementary tools.

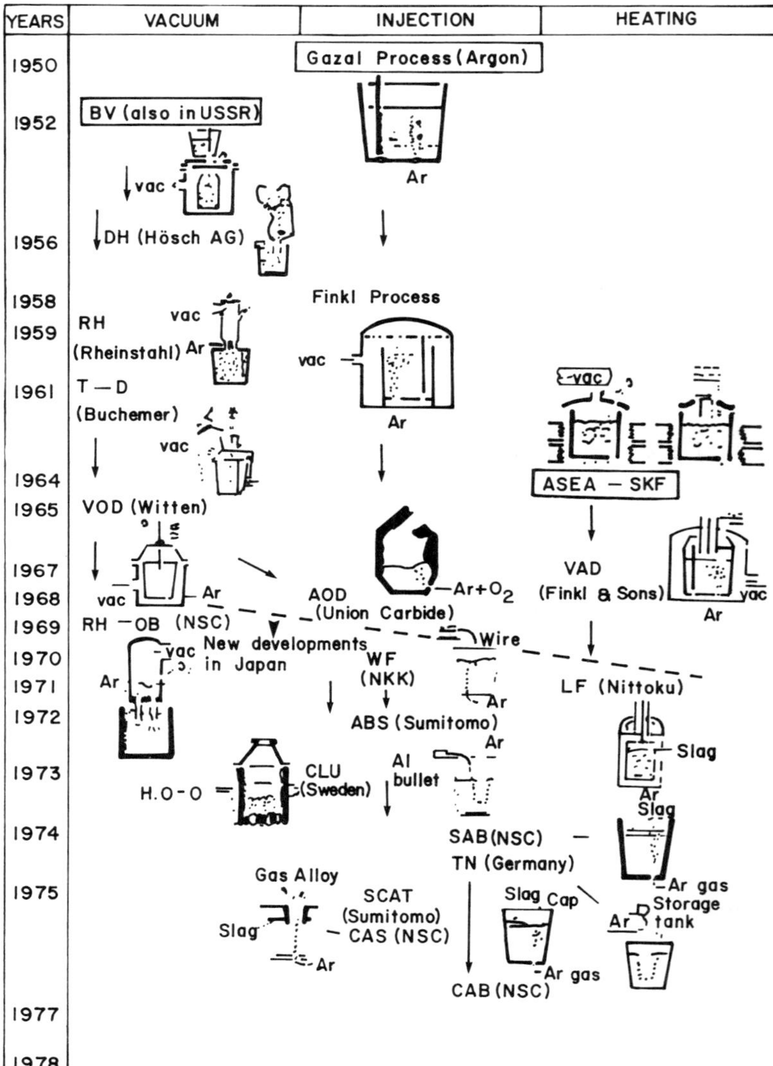

FIG. 6.1. Historical presentation of the evolution of ladle metallurgy operations.

Parallel to the industrial developments a great deal of useful research has been done on various aspects of ladle metallurgy. It is generally appreciated that, apart from process chemistry, the fluid flow phenomena, such as the flow patterns and the extent of agitation, play a key role in determining the overall efficiency of these operations. In general ladle metallurgical systems may be agitated by the injection of a gas stream, by vacuum-driven flows, and by the use of an electromagnetic force field. Within this framework we shall review recent efforts aimed at a better understanding of the behavior of these systems through the combined use of mathematical modeling, physical modeling, and plant-scale experimentation.

In considering the turbulent recirculating flows that are characteristic of these systems in general, from the standpoint of transport phenomena we are faced with three groups of problems:

1. The definition of the gross features of the circulation system, that is, the overall flow patterns and the mixing time
2. The definition of heat, mass, and momentum transfer between the melt and the bounding surfaces (e.g., problems concerned with refractory erosion, slag–metal reactions, and oxygen transfer from the refractory to the bath)
3. The description of the detailed structure of the turbulent flow, such as the map of the turbulent energy dissipation, which is necessary for modeling deoxidation kinetics; the coalescence of inclusion particles; and in general the interaction between suspended particles and the melt

In reviewing previous work, it is seen that major advances have been made regarding the definition of the overall flow patterns and the mixing times through the use of physical modeling and the application of various mathematical turbulence models.[5–13] Indeed, it may be stated with some confidence that these problems are well understood and that for certain geometries the estimation of the mixing time is essentially a routine task.

For example, Figure 6.2 shows a plot of the mixing time against the rate of energy dissipation (i.e., the total power input into the system). It is seen that most

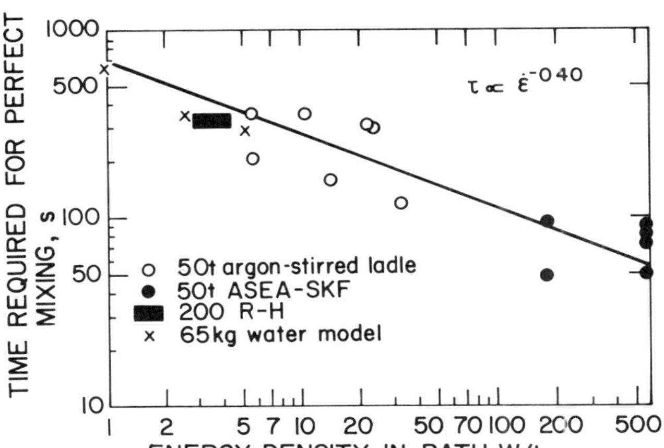

FIG. 6.2. Plot of the time required to attain perfect mixing against the energy density in the bath, after Nakanishi et al.[13]

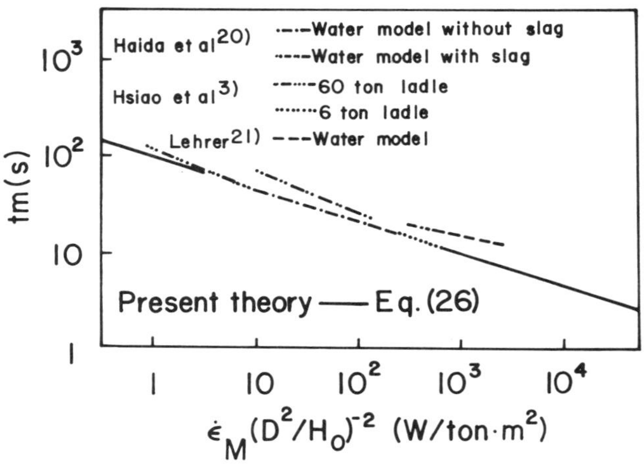

FIG. 6.3. Plot of the time required for perfect mixing as a function of the energy density in the bath, also showing the effect of system geometry, after Sand and Mori.[31]

systems of metallurgical interest may be reasonably well represented on this plot. A further generalization of this plot is shown in Figure 6.3, which incorporates the effect of the system geometry.

In this context mixing time is defined as the time required to obtain a uniform dispersion of an added tracer. While the mixing time is an imperfect measure because the term *uniform dispersion* is, of course, arbitrary, it has found considerable practical use for the rapid, if approximate, characterization of systems.

It should be noted that there are sound fundamental reasons why such an improved understanding of these systems was readily developed. As far as the circulation of these systems is concerned, in the bulk, the inertial terms represent the key factors; thus, the overall flow field may be readily modeled (either mathematically or physically) even if the viscous components (including the eddy phenomena) are not accurately represented. This is a key feature of body force–driven flows.

The situation is rather less satisfactory regarding the other two problem areas: fluid-to-wall (or slag metal) transport and the definition of the fine structure of the flow. Notwithstanding their obvious practical importance, these aspects of the problem have received much less attention; while attempts have been made to extend the theoretical treatment to the explicit representation of these phenomena, this work has to be regarded as largely tentative at this stage, because of the absence of hard corroborating evidence.

The purpose of this chapter is to report on recent research addressing these problems. The results of this research is presented under two headings: "Gas-Bubble-Driven Circulation Systems" and "Electromagnetically Driven Flows." Within these two categories, we shall address mathematical model development, physical modeling and the comparison of the measurements with predictions, and plant-scale measurements. We shall conclude with a discussion of the interplay among these three approaches and possible optimal strategies that may be adapted to tackling problems of this type.

6.1. GAS-BUBBLE-DRIVEN CIRCULATION SYSTEMS

6.1.1. Mathematical Modeling

In essence, the gas-bubble-driven circulation system, such as that sketched in Figure 6.4, represents an axisymmetrical turbulent recirculating flow problem of the type that has been tackled by numerous investigators.[5–8,10–12] One of the main problems encountered in previous studies was to obtain a proper representation of the boundary between the gas-bubble-rich jet cone region and the bulk of the liquid both in terms of position and in terms of stating the proper boundary conditions for the velocity and the velocity gradients.

The following assumptions have been made in the development of the fluid flow equations:

1. The two-phase region was confined to a plume, the dimensions of which were determined experimentally (e.g., from the visual observation of the "breakthrough zone."[8,10]
2. The two-phase region was assumed to be homogeneous.
3. The velocity and the momentum flux were assumed to be continuous across the jet cone (plume) boundary.
4. The two-dimensional turbulent Navier–Stokes equations were used in conjunction with the k–ε model to represent the fluid flow field, which was considered to be driven by the density difference in the two-phase region.

Within the framework of these assumptions the governing equations take the following form:†

Equation of continuity:

$$\frac{1}{r}\frac{\partial}{\partial r}(\rho r U_r) + \frac{\partial}{\partial z}(\rho U_z) = 0 \tag{1}$$

Momentum balance in the z direction:

$$\frac{1}{r}\frac{\partial}{\partial r}(\rho r U_r U_z) + \frac{\partial}{\partial z}(\rho U_z^2) = -\frac{\partial p}{\partial z} + \frac{1}{r}\frac{\partial}{\partial r}\left(r\mu_{\text{eff}}\frac{\partial U_z}{\partial r}\right) + 2\frac{\partial}{\partial z}\left(\mu_{\text{eff}}\frac{\partial U_z}{\partial z}\right) + \frac{1}{r}\frac{\partial}{\partial r}\left(\mu_{\text{eff}} r\frac{\partial U_r}{\partial z}\right) + \rho g\bar{\alpha} \tag{2}$$

†Figure 6.4 shows a definition of the coordinate system.

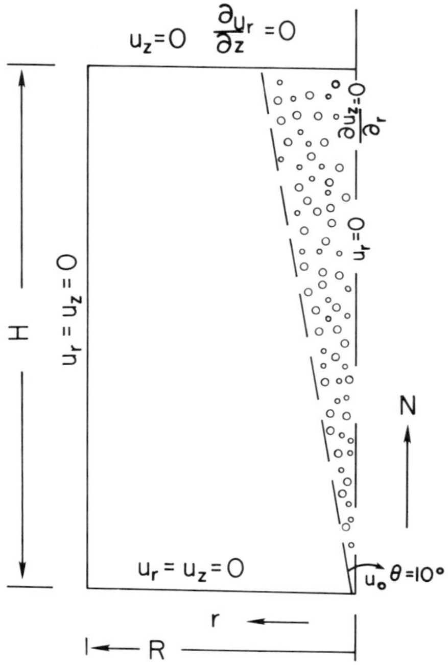

FIG. 6.4. Physical concept of the mathematical model of an argon-stirred ladle.

Momentum balance in the r direction:

$$\frac{1}{r}\frac{\partial}{\partial r}(\rho r U_r^2) + \frac{\partial}{\partial z}(\rho U_r U_z) = -\frac{\partial p}{\partial r} + \frac{2}{r}\frac{\partial}{\partial r}\left(r\mu_{\text{eff}}\frac{\partial U_r}{\partial r}\right) + \frac{\partial}{\partial z}\left(\mu_{\text{eff}}\frac{\partial U_r}{\partial z}\right) + \frac{\partial}{\partial z}\left(\mu_{\text{eff}}\frac{\partial U_z}{\partial r}\right) - \frac{2U_r\mu_{\text{eff}}}{r^2} \quad (3)$$

where

$$\rho = \rho_1, \qquad r > r_c \quad (4)$$

$$\rho = \bar{\alpha}\rho_g + (1 - \bar{\alpha})\rho, \qquad r < r_c \quad (5)$$

and $\bar{\alpha}$ is the void fraction of the gas.

It follows that for the purpose of modeling the whole domain has been treated as a homogeneous medium with a spatially variable density. Using a drift flux model, which allows the slip between the bubbles and the fluid, the quantity $\bar{\alpha}$ may be expressed as[14]

$$\bar{\alpha} = \frac{1}{2\pi}\frac{\rho_g - \pi r_c^2\bar{\alpha}(1 - \bar{\alpha})U_\infty}{\int_0^{r_c} rU_z\, dr} \quad (6)$$

Where U_∞ is the rising velocity of a characteristic single bubble, which is estimated as 40 cm/s.

The boundary conditions for the equation of motion take the following form:

At the axis (expressing symmetry):

$$U_r = 0, \qquad r = 0 \tag{7}$$

$$\partial U_z/\partial r = 0, \qquad r = 0 \tag{8}$$

At the walls (expressing zero velocity at the walls):

$$U_r = U_z = 0 \tag{9}$$

At the free surface (expressing zero shear at the free surface):

$$U_z = 0, \qquad z = H \tag{10}$$

$$\partial U_r/\partial z = 0, \qquad z = H \tag{11}$$

At the orifice:

$$U_z = U_0, \qquad r < r_0, \qquad z = 0$$

or

$$U_z = 0, \qquad r > r_0, \qquad z < 0 \tag{12}$$

Equations (1)–(12) constitute the complete statement of the fluid flow problem except for the definition of effective viscosity. In the present case we may use the well-established k–ε model for computing the turbulent viscosity:[15,16]

$$\mu_{\text{eff}} = \mu + \mu_t \tag{13}$$

with

$$\mu_t = \rho C_{\text{D}} k^2/\varepsilon \tag{14}$$

where μ_{eff} is the effective viscosity,
k is the turbulent kinetic energy,
ε is the rate of turbulent kinetic energy dissipation, and
C_{D} is a constant.

It should be noted that k and ε depend on both the nature of the flow and the location within the system. For a given situation the spatial distributions of k and ε must be obtained through the solution of auxiliary equations.

While Eqs. (1)–(14) do define the fluid flow problem, in most practical situations one wishes to calculate convective mass and heat transport in order to predict tracer dispersion–refining kinetics and the rate of dissolution of additives.

A. Mass Transfer

Using the shorthand of vector notation for convective mass transfer, we have

$$\frac{DC_R}{Dt} = \nabla D_{\text{eff}} \nabla C_{\text{R}} + \dot{r}_{\text{R}} \tag{15}$$

where $$\frac{D}{Dt}(\ ) = \frac{\partial}{\partial t}(\ ) + U_r \frac{\partial}{\partial r}(\ ) + U_z \frac{\partial}{\partial z}(\ ) \tag{16}$$

for axisymmetric cylindrical systems, which will be employed here,
C_{R} is the reactant or tracer concentration,
$D_{\text{eff}} = D_t + D$ is the effective diffusivity, and $\dot{r}_{\text{R}}$ is the rate of reaction that may produce or consume species in the bulk.

If tracer dispersion is being studied, $r_{\text{R}} = 0$. However, if we wish to model desulfurization, then C_{R} is the sulfur concentration, and the rate of reaction has to be considered.

Clearly, the mass balance equation [Eq. (15)] is closely related to the fluid flow problem, since the velocity field must be known before Eq. (15) may be solved. Furthermore, μ_t and D_t are related through

$$\frac{\mu_t}{\rho D_t} \simeq 1 \tag{17}$$

The boundary conditions for Eq. (15) is largely analogous to those given for the fluid flow equations, that is, zero gradient at the axis of symmetry and at the nonreacting walls.

If a reaction does take place at the slag–metal interface, it must be represented by an expression of the following type:

$$D \frac{\partial C_{\text{R}}}{\partial z} = f_{\text{R}}(C_{\text{R}}, T, \ldots) \tag{18}$$

where f_{R} designates the appropriate kinetic expression for the surface reaction.

B. Heat Transfer

In ladle metallurgy practice, heat transfer problems naturally arise in the consideration of heat losses, when the system operates as a ladle furnace with the addition of thermal energy via arcs, or when the melting dissolution of alloying additions is being investigated.

From the analogy with mass transfer, the heat transfer process may be represented by writing

$$\rho C_p \frac{DT}{Dt} = \nabla k_{\text{eff}} \nabla T + \dot{q} \tag{19}$$

where T is temperature,
C_p is specific heat,
k_{eff} is the effective thermal conductivity, and
$\dot{q}$ is the rate of heat generation or absorption per unit volume in the bulk.

On the analogy to Eq. (17), we have

$$\frac{C_p \mu_t}{k_t} \simeq 1 \tag{20}$$

The boundary conditions must express symmetry and the continuity of the heat flux and temperature at the boundary surfaces.

Equations (1)–(20), together with appropriate subsidiary relationships, such as the boundary conditions, the turbulence model, and the kinetic expressions, provide a complete statement of the problem, at least in principle. Thus, through the numerical solution of these equations, one may calculate

the velocity field,
the turbulent energy distribution, and
the map of the turbulent energy dissipation,

and hence tracer dispersion (homogenization) rates and mass transfer rates, such as desulfurization and deoxidation kinetics, the temperature profiles, and the heat transfer rates.

As a practical matter, in view of the many assumptions made *the results of all these calculations will have to be regarded as merely estimates* unless some experimental verification is obtainable.

Once verified experimentally, the predictions based on the model may then be treated with confidence. This is a very important point which illustrates the clear interdependence of mathematical modeling and the experimental measurements.

6.1.2. Physical Modeling

The gas-bubble-stirred ladles are quite complex, two-phase flow systems, the rigorous physical modeling of which would be very difficult. While one can achieve geometric similarity in the sense of proportioning the physical dimensions of the apparatus, it is not clear whether the gas plume dimensions could be accurately reproduced. Furthermore, bubble formation, breakup, and coalescence are likely to be quite different in aqueous and metallic systems.

By the same token it would be extremely difficult to develop a rigorous physical model for slag–metal reactions or the reaction of the melt with injected solid particles.

Based on the discussion in Chapter 3, the most appropriate use of physical modeling in this context would be (1) to conduct ad hoc measurements, and (2) to measure certain physical phenomena in the system with the view of verifying the mathematical models.

A. Ad Hoc Measurements

In terms of ad hoc measurements it may be worthwhile to construct a geometrically similar system, for example, one made of Plexiglas and employing air and water as the working fluids. While the actual scaling need not be a crucial consideration, it is important to ensure that the scale selected is large enough so that wall effects do not become spuriously dominant and that one is in the proper flow regime (laminar or turbulent).

Then it may be desirable to ascertain that for the injection rates in the model and in the prototype one is in the same regime (i.e., discrete bubble formation, jetting, and the like). In addition one would wish to ensure that the ratio (system volume)/(volumetric gas flow rate) is comparable in the model and in the prototype. Then one may do tracer dispersion tests to examine the optimal location of the injection point, identify the stagnant regions, and the like. These data should also provide some information on the mixing time, although the quantitative interpretation of these measurements may not be straightforward.

B. Quantitative Measurements

In making quantitative measurements, one would follow the same guidelines as described above to ensure correspondence between the real system and the model. However, the principal objective of these studies is to provide *a critical test of the mathematical model* and if necessary to develop some form of information that may be incorporated into the model. The example that follows illustrates these considerations.

Consider an apparatus for studying bubble-driven circulation systems that is constructed to allow the convenient measurement of the velocity fields, turbulence characteristics, and heat transfer caused by the melting of immersed ice rods in a cylindrical tank containing water, which is being agitated by an ascending gas-bubble stream. The gas stream is introduced through the bottom of the container via a centrally located orifice. The tank is surrounded by a square tank, also filled with water, in order to eliminate the parallaxis effects of the optical measurements.

Figure 6.5 shows a sketch of the apparatus, which also indicates the principal dimensions of the container and the supporting and positioning assembly used

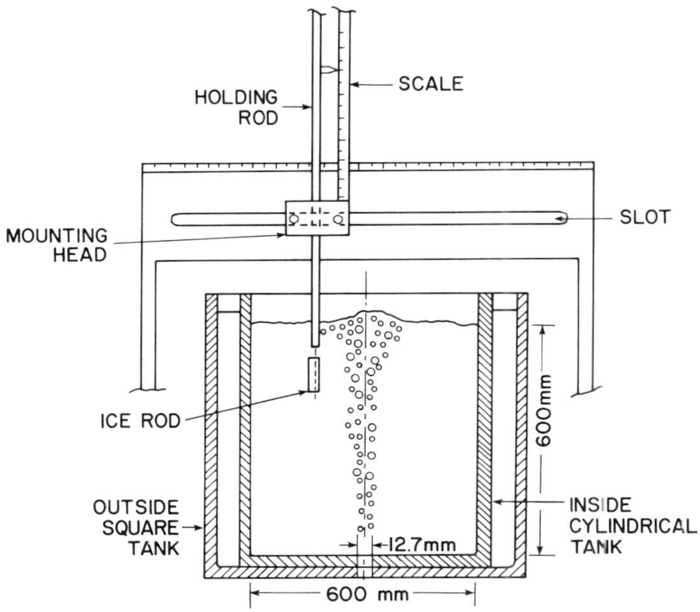

FIG. 6.5. Schematic sketch of the experimental arrangement for the physical model.[10,34]

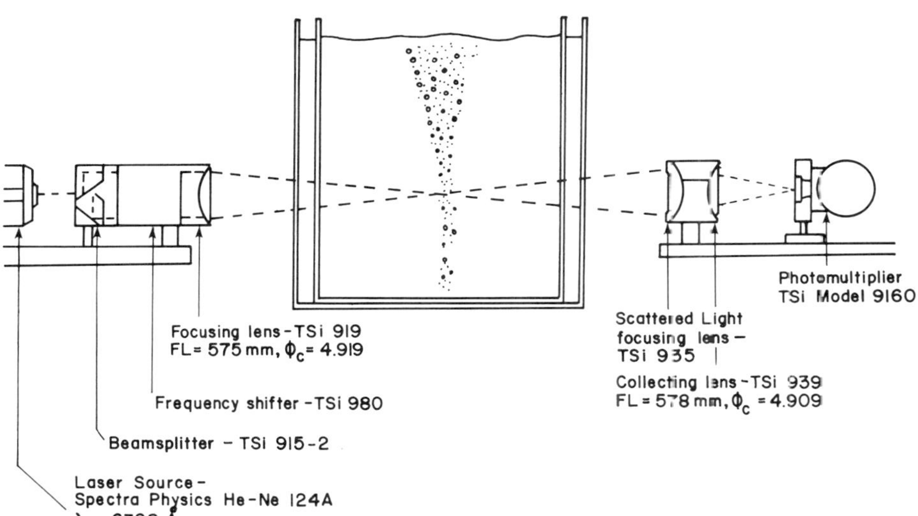

FIG. 6.6. Arrangement for the laser velocimeter.[10,34]

for the melting of the ice rods. The actual melting rate is determined using photographic techniques. A schematic outline of the apparatus as a whole is shown in Figure 6.6, where it is seen that the measurements of the velocities and of the turbulence characteristics is accomplished through the use of a laser–Doppler velocimeter employing a He-Ne laser. The velocimeter is operated in a dual-beam mode, with a analysis of the forward-scattered light.

Quantities derived from the measurements are the time-averaged velocity component, the turbulent kinetic energy (k), the rate of turbulent energy dissipation (ε), the Reynolds stresses, the turbulence intensity, and the like. The procedure for obtaining Reynolds stresses by such means is well documented.[17]

C. Experimental Measurements and Computed Results

This section presents a selection of the experimental measurements, which will be compared with the computed values of the corresponding parameters. The data include information on the velocities, on the turbulent kinetic energy, and on the Reynolds stresses.

1. MEAN VELOCITY FIELD. Figure 6.7 shows a comparison between the experimentally measured and the theoretically predicted velocity fields. It is seen that the agreement between measurements and predictions is reasonably good. It should be noted that the use of k–ε model in conjunction with appropriate wall functions gives a significantly better agreement between the measurements and the predictions, especially in the near wall regions, than does the use of simpler

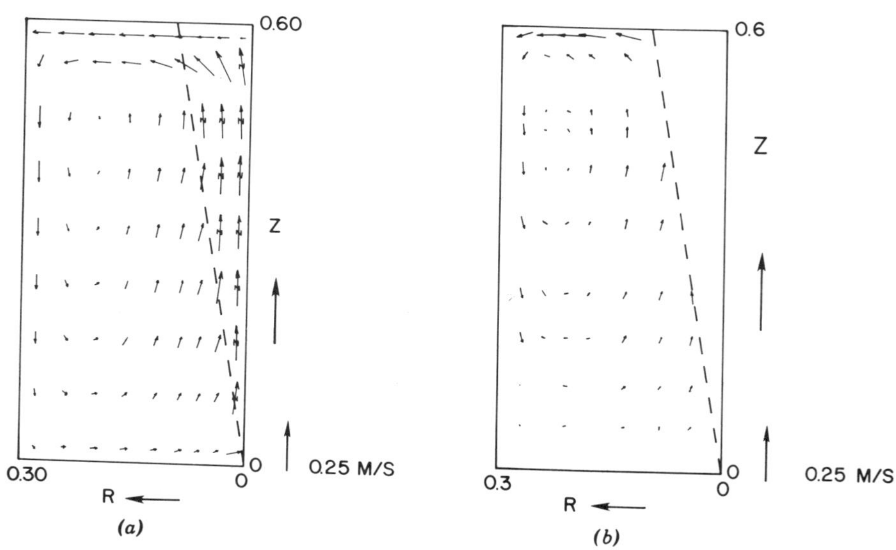

FIG. 6.7. Comparison of the experimentally measured and theoretically predicted velocity fields for the water model. (*a*) Experimental measurements. (*b*) Theoretical predictions.[10,34]

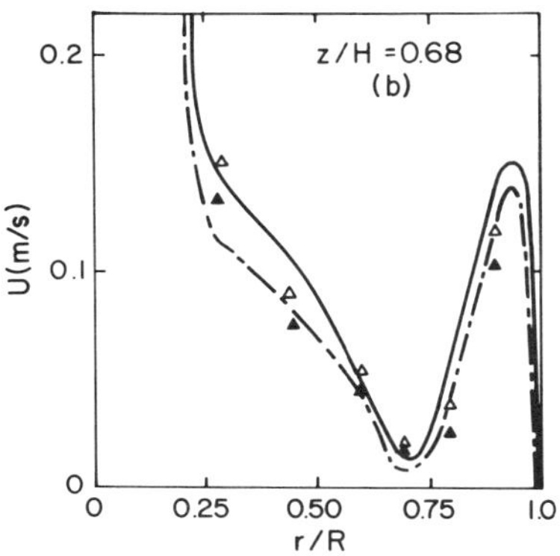

FIG. 6.8. Comparison of the theoretically predicted (continuous lines) and experimentally measured (broken lines) velocities for two linear gas velocities (▲, 1.6 m/s; △, 3.2 m/s).[10,34]

computational proceedings, employing a single value of the effective viscosity.[10] The wall functions specify the velocity profile in the vicinity of the wall. This type of relationship has been deduced semiempirically.[15]

Figure 6.8 shows a comparison between predicted and measured absolute values of the velocity vector at a particular axial position. It is seen that the agreement is quite reasonable throughout the domain.

2. Turbulence Characteristics of the System. Figure 6.9 shows the profiles of the root mean square (rms) velocity components normalized with respect to the mean velocity in the jet cone at the corresponding vertical position. A somewhat simplified analytical method was used for estimating the latter quantity.[11] The turbulence appears to be fairly isotropic except in the vicinity of the solid surfaces (side walls and the bottom), which is essentially in line with expectations.

Figures 6.10*a* and 6.10*b* show a comparison between the experimentally measured and the calculated values of the turbulent kinetic energy. The agreement seems to be within about a factor of 2 in most cases, which is not unreasonable, but this discrepancy indicates that further study of these pheno-

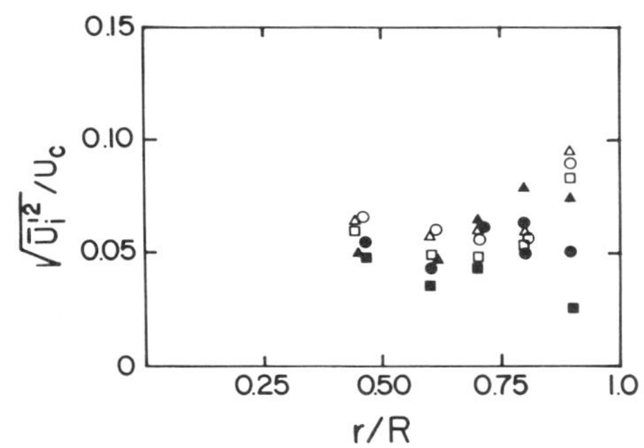

FIG. 6.9. Radial variation of the rms velocities for an orifice velocity of 1.6 m/s. [△, $(U_z'^2/U_c)^{1/2}$; ■ $(U_r'^2/U_c)^{1/2}$; ○ $(U_\theta'^2/U_c)^{1/2}$].[10,34]

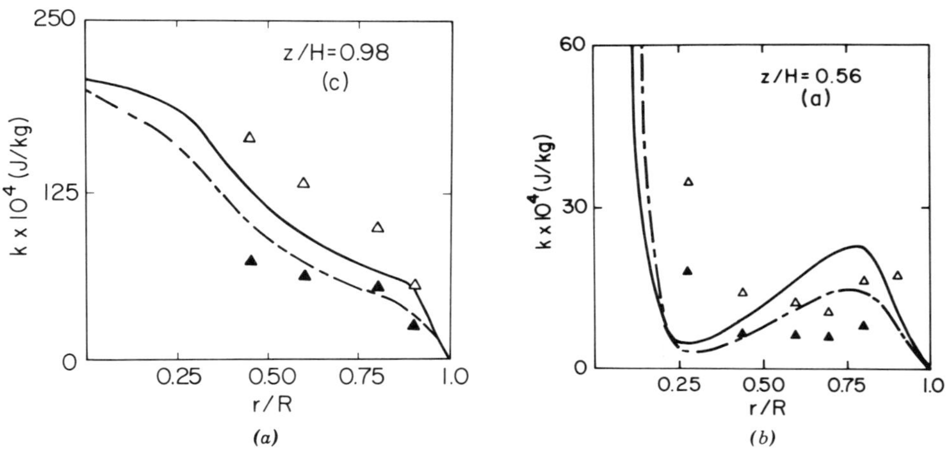

FIG. 6.10. Comparison of the measurements and predictions for the radial variation of the turbulent kinetic energy at two vertical positions (△, 1.62 m/s; ▲, 3.2 m/s).[10,34]

mena is justified. Such further theoretical work could involve either the adjustment of the parameters in the $k-\varepsilon$ model or perhaps the development of an alternative modeling approach.

The turbulent kinetic energy has much higher values at the free surface than in the bulk. This behavior underlines the fact that Levich's ideas of turbulence damping by surface tension forces would be inappropriate for such highly turbulent fluid flows, where the surface is disrupted by waves and breaking bubbles. This finding does, of course, have important implications regarding mass transfer phenomena.[18]

Figure 6.11 shows the computed contours of the ratio of turbulence viscosity to molecular viscosity. It is seen that this ratio is quite high, indicating that the flow is highly turbulent. Furthermore, substantial variations exist in the turbulent viscosity, so the use of a single constant value is unlikely to be satisfactory, except as a first approximation.

3. Heat Transfer and Phase Change. In this series of experiments, the rate at which ice rods melt upon immersion into a gas-bubble-agitated water pool is studied. The main motivation for this work is to obtain better insight into the effect of the velocity fields and the local turbulence parameters in the system on heat transfer between the fluid and an immersed solid body. These types of problems are of considerable practical interest both regarding wall erosion and dissolution kinetics.

Figure 6.12 shows typical plots of the position of the melting interface against time obtained within the jet cone and outside the jet cone. The near linearity of these plots indicates that the time required for the attainment of steady-state melting conditions is relatively short.

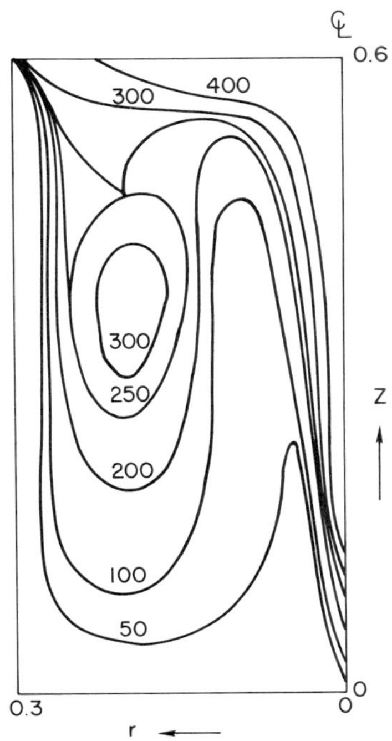

FIG. 6.11. Computed contours of the ratio of effective viscosity to molecular viscosity for a linear air velocity of 1.6 m/s at the orifice.[10,34]

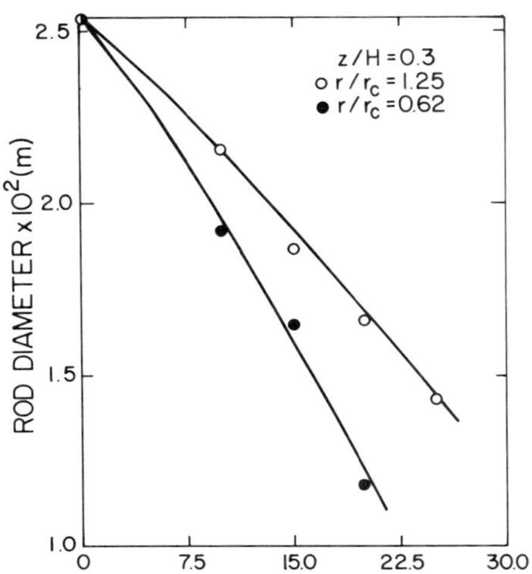

FIG. 6.12. Rate of dissolution of an ice rod immersed in an agitated pool as a function of radial position.[11]

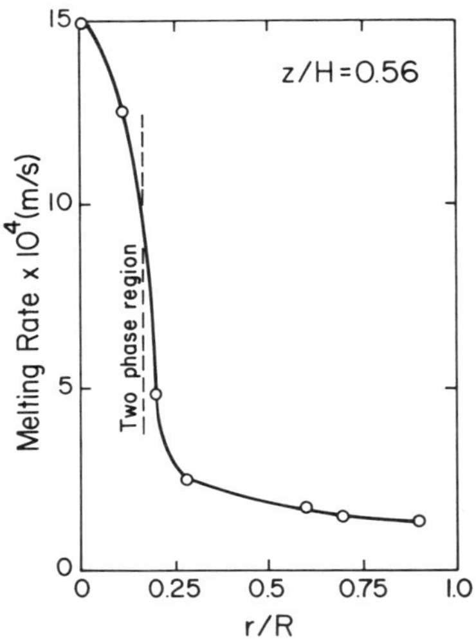

FIG. 6.13. Radial distribution of the initial rate of dissolution, indicating much faster rates in the central region.[11]

Figure 6.13 shows the radial variation in the melting rate of a constant axial position. It is seen that the melting rate is the highest in the jet cone and that it decreases rapidly with the radial distance from the axis of the gas jet.

Figure 6.14 summarizes the radial variation in the melting rates at various axial positions in the jet cone. The maximum melting rate is seen to occur at the center at an intermediate height. It may be shown that the fluid velocity is the maximum at this position.

Measurements show the melting vapor to be high at the surface of the pool, where the turbulence is high, and to be low at the mid-radius and mid-height, corresponding to the center of the vortex of recirculation. These measurements may be represented by the following correlation:

$$N_{\mathrm{Nu}} = 0.388(N_{\mathrm{Re}}\, Tu)^{0.8} N_{\mathrm{Pr}}^{1/3} \tag{21}$$

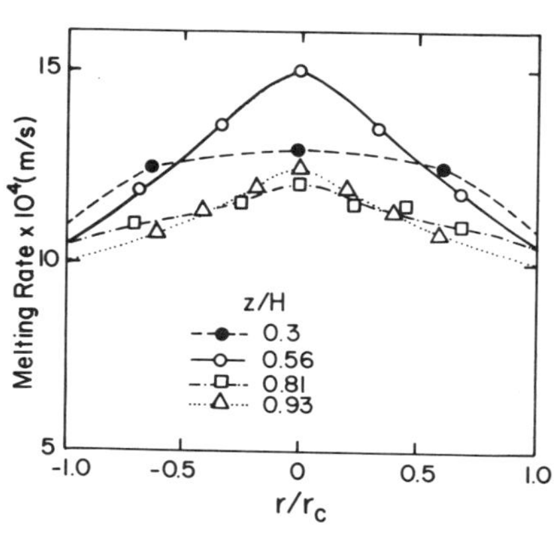

FIG. 6.14. Radial distribution of the initial melting rate in the jet cone at different axial position.[11]

for

$$100 = N_{\mathrm{Re}} \geqslant 2000$$

The Prandtl number is not varied in these measurements. Here

$N_{\mathrm{Nu}} = hL/k$ is the Nusselt number,
$N_{\mathrm{Re}} = UL\rho/\mu$ is the Reynolds number,
$Tu = (U')_{\mathrm{rms}}/U_{\mathrm{c}}$ is the turbulence intensity,
$N_{\mathrm{Pr}} = C_p\mu/k$ is the Prandtl number, and
U_{c} = a reference velocity at the centerline.

The important finding reflected by Eq. (21) is the fact that the local fluid-to-surface heat transfer coefficient depends not only on the Reynolds number but also on the local turbulence intensity.

At this stage it may be worthwhile to summarize the results of the physical modeling work described here. Measurements were carried out to determine the velocity fields and the spatial distribution of the turbulent kinetic energy and of certain turbulence characteristics. On comparing these data with the theoretical predictions, based on the mathematical model very good agreement was found regarding the velocity and reasonable agreement was obtained for the turbulent kinetic energy distribution. Turbulence was found to be isotropic in the bulk, but significant anistropy was seen to exist in the vicinity of the walls.

It follows that these measurements have proven the appropriateness of the model as far as fluid flow and turbulence are concentrated in the bulk. Since the bulk region is a single-phase system, it is reasonable to expect the mathematical model to be valid for predicting velocity fields and turbulence levels in molten steel systems as well.

The measurements obtained for the local heat transfer rates come under a different category. The near-wall conditions, and hence fluid-to-wall heat or mass transfer, cannot be readily predicted from first principles; indeed, one does need measurements to come up with a realistic representation of these important aspects of the problem.

The correlation represented by Eq. (21) should be helpful, not only for estimating the local heat or mass transfer coefficients for a variety of systems, but also for providing realistic values for the boundary conditions describing the local fluid to wall heat flux.

6.1.3. Plant-Scale Measurements

Ultimately, the test of any model is how well it predicts the behavior of real systems. Indeed, such tests are essential before any model is put to widespread use or employed in a critical application.

At present there are only a limited number of plant-scale measurements for which sufficient details are available on the system to allow a meaningful comparison between the measurements and the predictions. Nonetheless, it is important to cite them because they provide the ultimate test of the results obtained in the mathematical and water modeling work described here.

A. Mixing

Mixing is the topic that has been most extensively studied. Numerous data are available on tracer dispersion and mixing in vessels of various sizes.[8,10,13] Since this field is now quite well understood, we shall present only one typical example illustrating that the theoretical predictions provide satisfactory answers regarding mixing time. In most industrial scale systems, where the flow is reasonably turbulent, the dispersion of the tracer takes place by two mechanisms, bulk convection and eddy diffusion, as sketched in Figure 6.15. The actual convective time scale is very short in these systems, typically of the order of a few seconds (linear dimensions being of the order of 1–2 m and the characteristic velocity being about 0.3 m/s). Thus, the rate-determining step in tracer dispersion or mixing is eddy diffusion between the respective streamlines. This behavior is consistent with the earlier discussion of the fact that the mixing characteristics of a system may be readily related to the total energy input.

DESULFURIZATION KINETICS. Figure 6.16 is a sketch of a typical arrangment used for the desulfurization of steel melts.[18] A solid desulfurization agent is injected into the melt, together with a carrier gas, through a suitably placed lance. The overall rate of the desulfurization reactions depend on two factors in addition to the obvious criteria determined by thermodynamic equilibria:

1. The rate at which the individual particles react is determined by the Kolmorgoroff eddies,[19,21] which clearly depend on the local rate of turbulent energy dissipation.
2. In connection with mixing, the rate at which the reactant—that is, the sulfur—is transported to the regions where the reaction is actually taking place is limited by eddy diffusion.

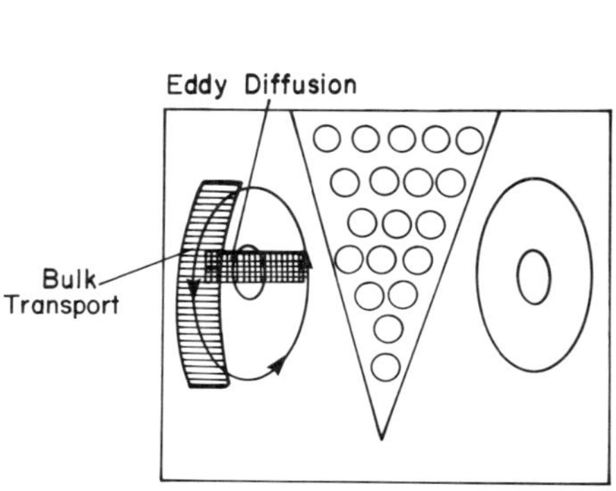

FIG. 6.15. Mechanism of mixing in gas-stirred systems by bulk transport and eddy diffusion.

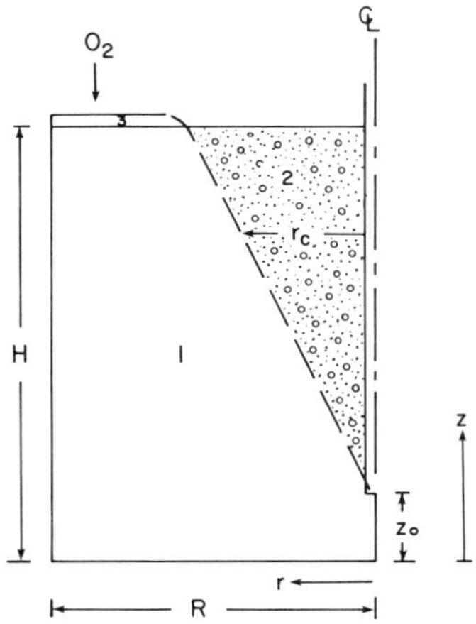

FIG. 6.16. Schematic representation of the desulfurization model (1, molten steel; 2, three-phase zone; 3, slag layer).

Figures 6.17*a* and 6.17*b* show a comparison between the experimentally measured and theoretically predicted desulfurization rates for a pilot-scale and an industrial-scale system, respectively.[18] While these two sets of curves appear to be similar, the actual reaction mechanisms differ substantially.

In the pilot-scale operation, where the system was heavily deoxidized, the reaction took place in the jet plume, while in the industrial-scale unit, which was not strongly deoxidized, most of the desulfurization took place at the slag–metal interface. The very good mixing produced by the gas bubbles in the large-scale system provided an adequate rate of transport to the reaction zone in both cases.

B. Heat or Mass Transfer to Immersed Surfaces

A key problem in the operation of ladle metallurgical operations is the definition of the factors that govern wall erosion, and hence the lining life. Relatively little modeling work has been done in this area, but a useful start has been made regarding the study of the rate at which immersed bodies dissolve.

Experiments have been performed at MEFOS, in Lulea, Sweden, on the solution of graphite rods in a 6-ton iron melt. The experimental results are given in Figure 6.18 in the form of the local mass transfer coefficients.

The numerical values of these mass transfer coefficients would be difficult to predict from first principles (the same applies to the wall erosion rates) because the local fluxes at the solid surface depend not only on the local velocities but also on the turbulence levels in the system. However, a preliminary prediction may be made regarding the numerical values of these transfer coefficients through the appropriate scaling of water-modeling experiments, which reported that the conditions of turbulence were comparable in the water-model system and the pilot-scale experiments (the water models were designed with this objective in mind). The mass transfer coefficients may be scaled with the aid of the previously given formula:

$$N_{\mathrm{Nu}} = 0.388\,(N_{\mathrm{Re}}\,Tu)^{0.8}\,N_{\mathrm{Pr}}^{1/3}, \qquad 100\,N_{\mathrm{Re}} < 2000 \tag{22}$$

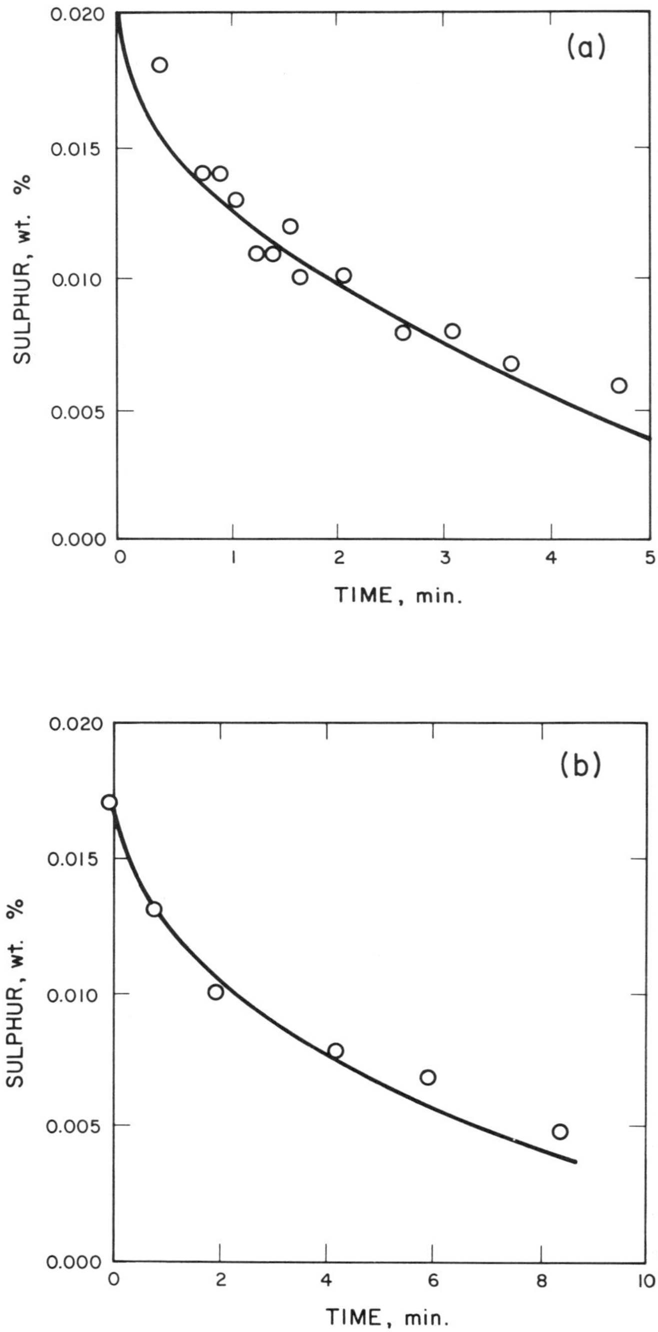

FIG. 6.17. Comparison of the experimentally measured and theoretically predicted sulfur content as a function of time for (*a*) a 6-ton heat, and (*b*) a 40-ton heat.[18]

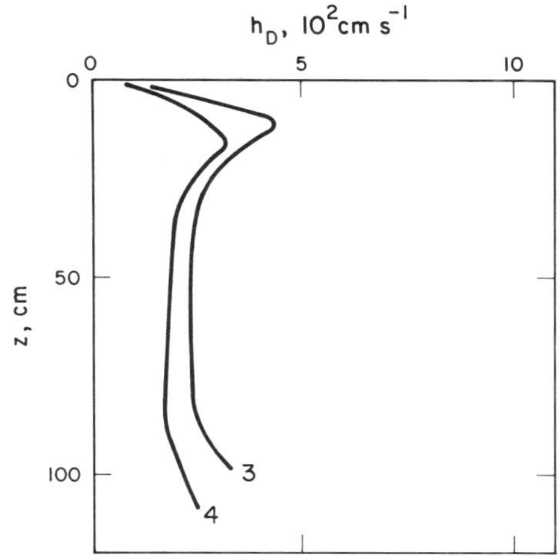

FIG. 6.18. Experimentally measured values of the mass transfer coefficient for the dissolution of a graphite electrode immersed in a 6-ton steel melt, showing the effect of immersion depth at two radial locations.

6.2. ELECTROMAGNETICALLY DRIVEN FLOWS

While the use of bubble stirring is perhaps the simplest way of agitating a volume of liquid metal, it is also possible to bring about such agitation by employing electromagnetic forces. In ladle metallurgy, such agitation is provided in the ASEA-SKF furnace and is also present in conventional induction furnaces. Metal motion brought about by electromagnetic forces is important in a number of other metal processing operations, such as the Hall–Héroult cell, used to produce aluminum (see Chapter 8) in electroslag refining and plasma processing. Here we present a general treatment of electromagnetically driven flows before examining work of relevance to ladle metallurgy.

6.2.1. Mathematical Modeling

Let us consider a typical induction furnace, such as sketched in Figure 6.19. An ac current is passed through the induction coils, and as a result current is induced in the melt contained in the furnace. This induced current performs a dual function:

1. It provides joule heating due to the J^2R term.
2. The interaction between the induced current and the magnetic field generated gives rise to a Lorentz force or electromagnetic force, which is responsible for driving the flow.

The equations describing the motion of a fluid under the influence of

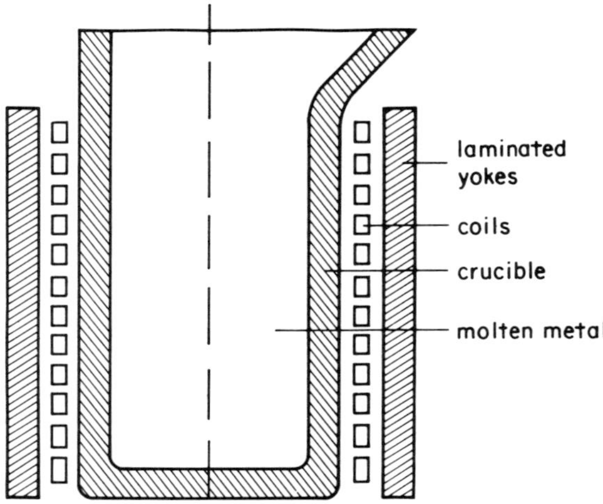

FIG. 6.19. Schematic sketch of an induction furnace.

electromagnetic stirring are, for laminar flow (assuming an incompressible fluid)

The continuity equation:

$$\nabla \cdot \mathbf{u} = 0 \tag{23}$$

The momentum balance equations (assuming a Newtonian fluid):

$$\rho \frac{D\mathbf{u}}{Dt} = -\nabla p + \mu \nabla^2 \mathbf{u} + \rho \mathbf{g} + \mathbf{F} \tag{24}$$

The turbulent flow equations:

$$\nabla \cdot \bar{\mathbf{u}} \tag{25}$$

$$\rho \frac{D\bar{\mathbf{u}}}{Dt} = -\nabla p + \nabla \cdot \mu_e \nabla \bar{\mathbf{u}} + \rho \mathbf{g} + \mathbf{F} \tag{26}$$

where v is the velocity vector,
ρ is the fluid density,
μ is the fluid viscosity,
p is the pressure,
$\mathbf{g}$ is the gravitational force vector,
$\mathbf{F}$ is the electromagnetic force vector,
$\bar{\mathbf{u}}$ is the time-averaged velocity vector, and
μ_e is the effective viscosity of the turbulent fluid.

Equation (26) may be solved with the aid of the $k-\varepsilon$ model, mentioned in Section 6.1.2.A.

Equations (23)–(26), together with the associated boundary conditions, represent the statement of the electromagnetically driven fluid flow problem. In the majority of cases one may proceed by solving Maxwell's equations first, in order to obtain the electromagnetic force field, and then using this information to compute the fluid flow field and, if applicable, the turbulence parameters of the system. Once the fluid flow field is known, one may readily extend the solution to a range of convective heat flow and mass transfer problems.

As a practical matter, these boundary conditions will tend to specify zero velocities at the solid surfaces; zero shear at the free surface; and, if applicable, symmetry along the same lines as in gas-bubble-stirred systems.

During the past decade a large number of computed results have been published concerning a range of applications, including induction furnaces, aluminum electrolysis, and electroslag refining.[22-30]

Figure 6.20 shows the computed velocity field for a 50-ton ASEA-SKF furnace, and Figure 6.21 shows computed streamline plots for a 12-ton melt for

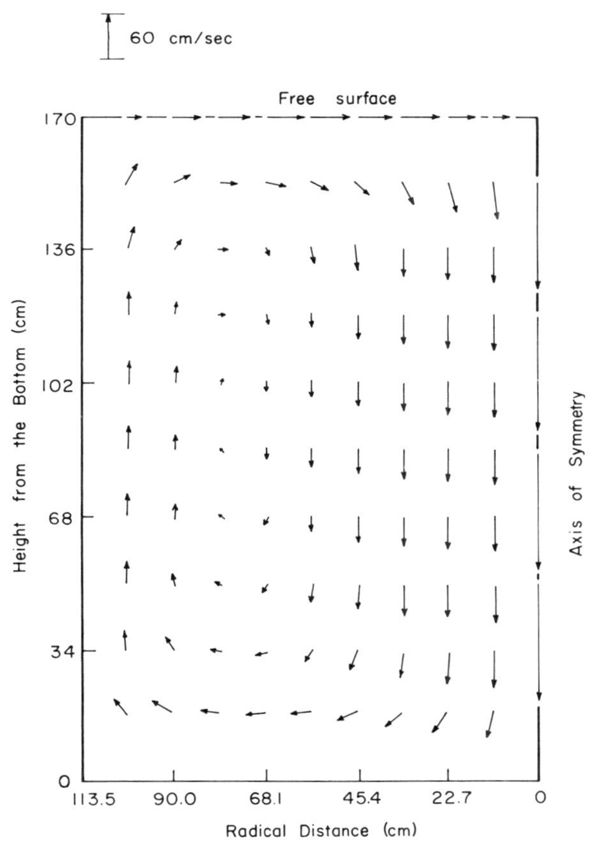

FIG. 6.20. Computed map of the velocity field in an ASEA-SKF furnace, after Nakanishi et al.[13]

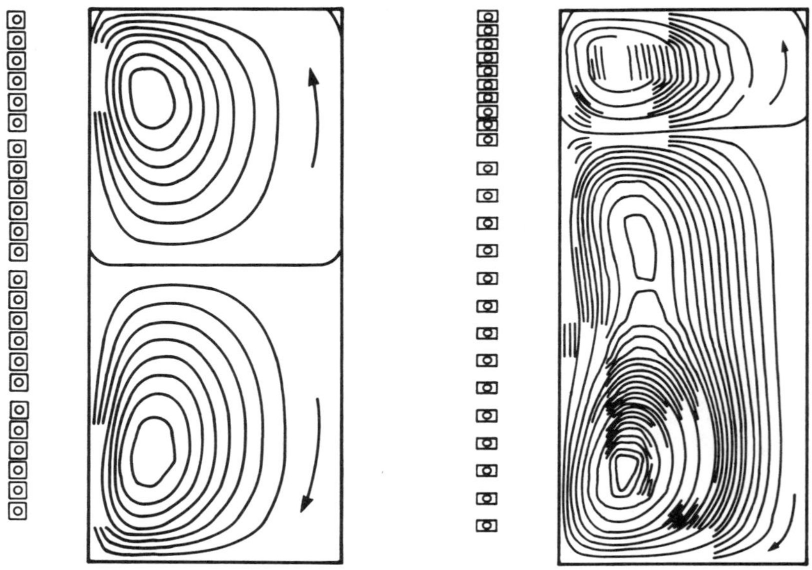

(a) (b)

FIG. 6.21. The computed streamline patterns for an induction furnace, after Tarapore et al.[28] (a) 3900-A current for a 12-ton melt. (b) 1000-A current an irregularly placed coils.

various values of the coil current and for various coil configurations.

The mathematical modeling of electromagnetically driven flows in metals processing is a nonroutine application of the transport equations, and for this reason it is of considerable interest to compare the theoretical predictions with the results of measurements. Fortunately, during the past decade a useful range of experimental measurements has been reported concerning both laboratory- and industrial-scale systems.

6.2.2. Physical Modeling

A. Laboratory-Scale Studies

In the physical modeling of fluid flow phenomena in induction furnaces, one is restricted in the choice of working fluids to good conductors of electricity on the one hand and to relatively low-temperature systems on the other, so that measurements may be readily carried out.

As a practical matter, most of the work done in this area has employed either mercury or Wood's metal as the working fluid. In this area early, interesting work has been reported by Tarapore and Evans, who measured the surface

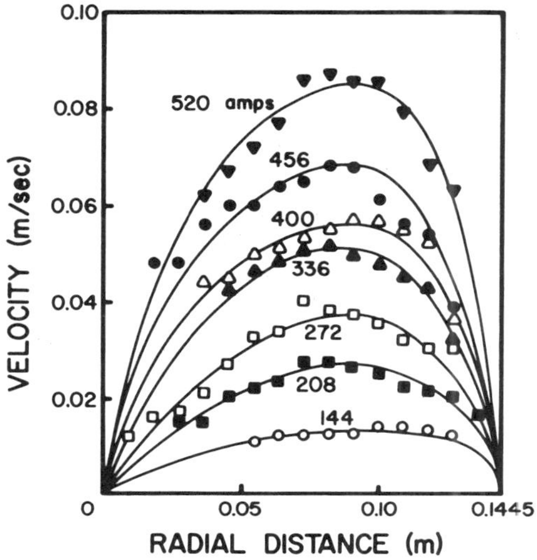

FIG. 6.22. Surface velocity profiles for a mercury pool, as measured by Tarapore and Evans.[22]

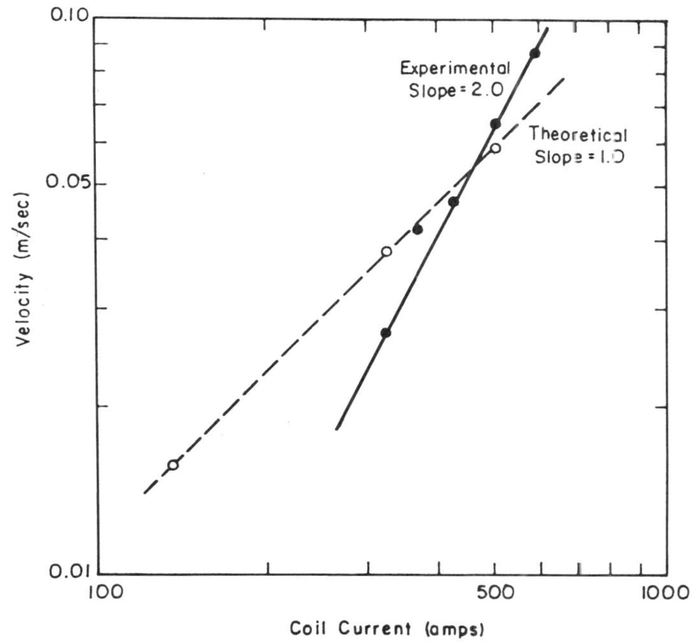

FIG. 6.23. Comparison of the experimentally measured and theoretically predicted maximum velocities for an inductively stirred mercury pool, after Tarapore and Evans.[22]

velocity of an inductively stirred mercury pool.[22] As seen in Figures 6.22 and 6.23, the agreement between measurements and predictions was quite reasonable.

Figure 6.24 shows a comparison between theoretical predictions and the experimentally measured velocity field in an inductively stirred pool of molten woods metal, as reported by Szekely et al.[23]

Perhaps the most careful measurements of melt velocities in models of induction furnaces has been reported by Moore and Hunt.[24] Figures 6.25 and 6.26 show a comparison of these measurements with predictions developed by

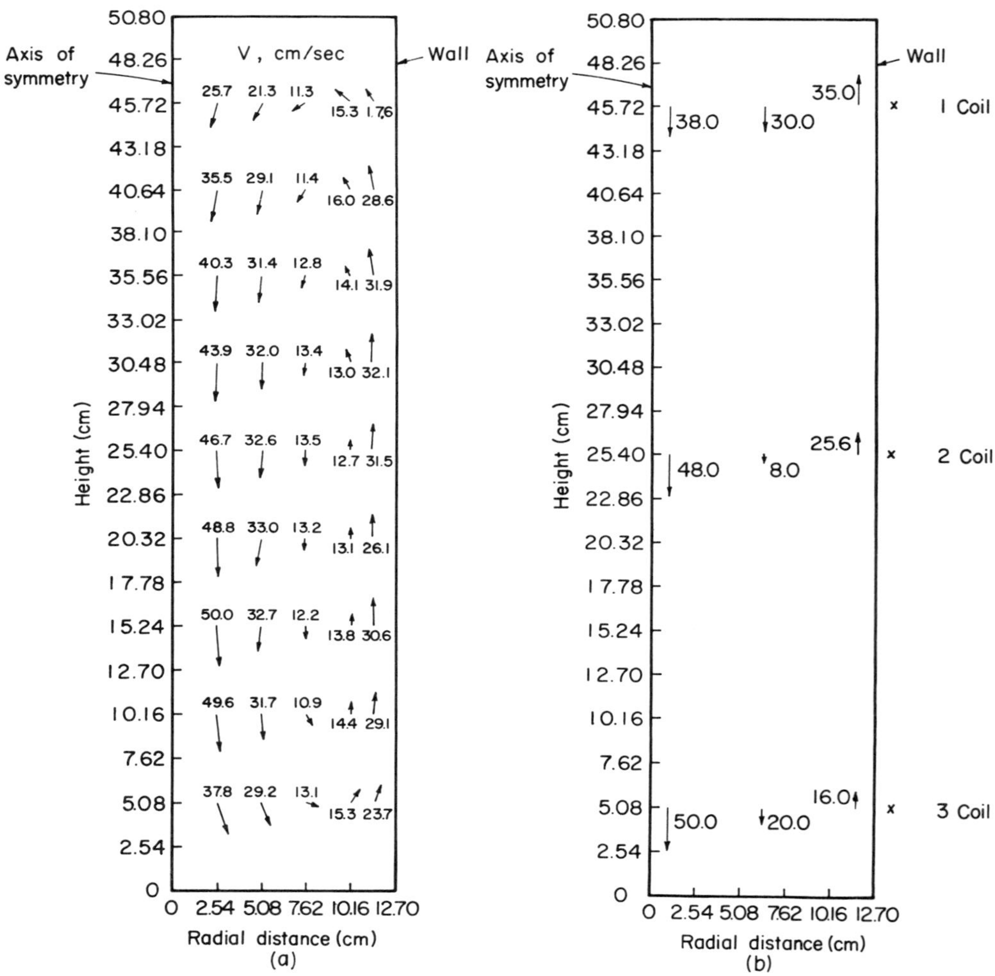

FIG. 6.24. Comparison of the theoretically predicted (right-hand plot) and experimentally measured (left-hand plot) velocities in an inductively stirred low-melting alloy, after Szekely et al.[23]

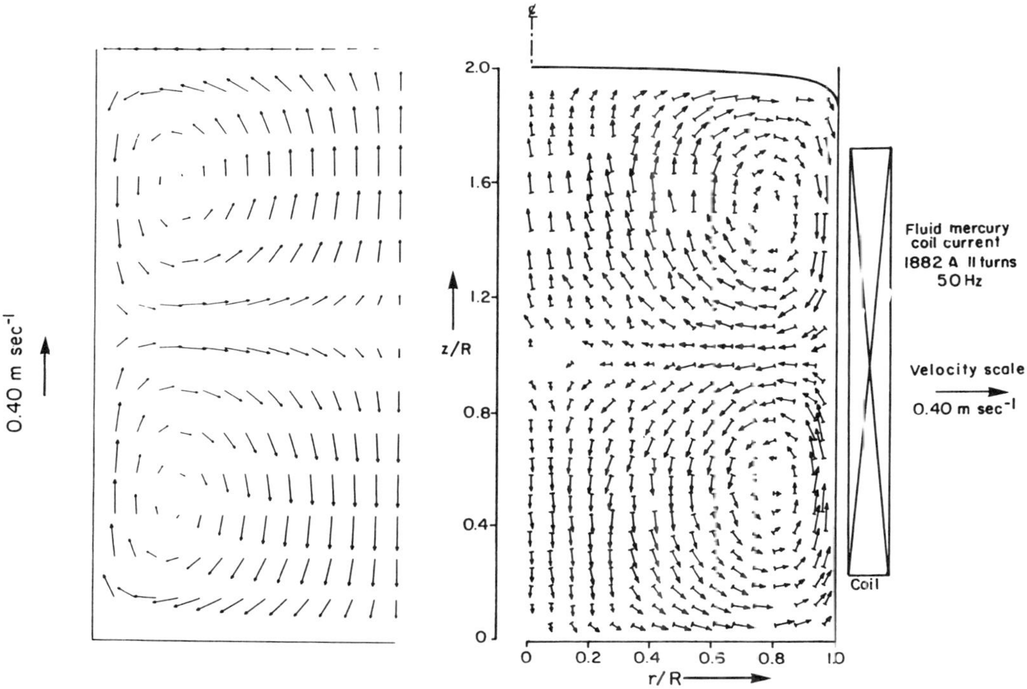

FIG. 6.25. Comparison of experimentally measured (right-hand plot) and theoretically predicted (left-hand plot) velocities in an inductively stirred mercury melt (measurements by Moore and Hunt,[24] and calculations by Evans).

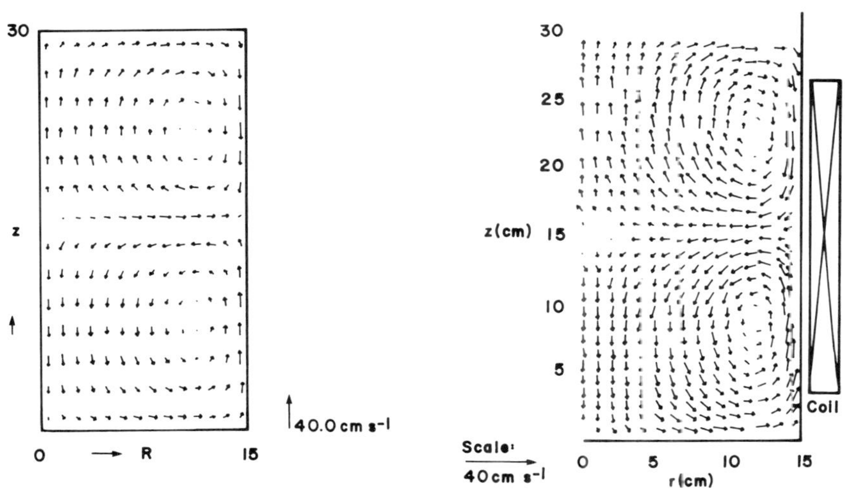

FIG. 6.26. Comparison of (*a*) experimentally measured velocity fields by Moore and Hunt[24] and (*b*) theoretical predictions by El-Kaddah and Szekely.[26]

Evans and Lympany,[25] and El-Kaddah and Szekely,[26] respectively. The essential quantitative agreement between these careful measurements and predictions is readily apparent.

B. Measurements at High Temperatures Including Plant- and Pilot-Plant-Scale Data

The quantitative determination of velocities and turbulence parameters in real high-temperature molten metal systems such as aluminum or steel poses major difficulties, although some interesting data have been reported in the literature in these areas, notably by Szekely et al.[27,29] Tarapore et al.,[28] and Tracy.[33]

Perhaps the easiest way of estimating melt velocities in high-temperature systems is by photographing the top free surface. Tables 6.1 and 6.2 show a comparison between measurements and predictions for steel melts, as reported by Szekely et al.[27,29] and Tracy.[33] These comparisons indicate reasonable validations of the model for industrials-scale systems. Figure 6.27 shows the results of Tarapore et al.[28] in a graphical form, again indicating good agreement between measurements and predictions. Figure 6.28 shows a com-

TABLE 6.1. Comparison of Experimentally Measured and Theoretically Predicted Surface Velocities for an Inductively Stirred Chromium–Nickel Melt

	Measured	Computed
Average surface velocity (cm/s, 500 kW)	18.19 ± 3.55	17.6
Average surface velocity (cm/s)	39.56 ± 7.69	32.7
Ratio of average surface velocities (2000 KW/500 KW)	2.17	1.86

SOURCE: From Szekely et al.[27]

TABLE 6.2. Comparison of Experimentally Measured and Theoretically Predicted Surface Velocities in a Laboratory Scale Induction Furnace

Current in Coil	Experimental Average Surface Velocity	Predicted Velocity
640	15	
800	22	22
952	29	27
1048	33	29

SOURCE: From Tracy.[33]

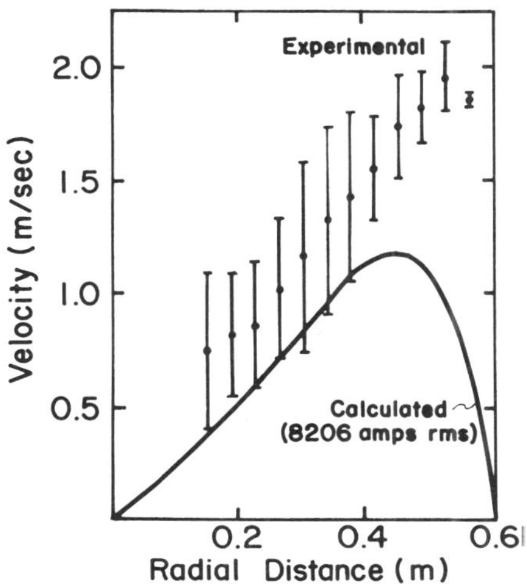

FIG. 6.27. Surface velocity measurements reported by Tarapore et al. for an industrial scale induction furnace. The data show the linear dependence of the velocity on the coil current.[28]

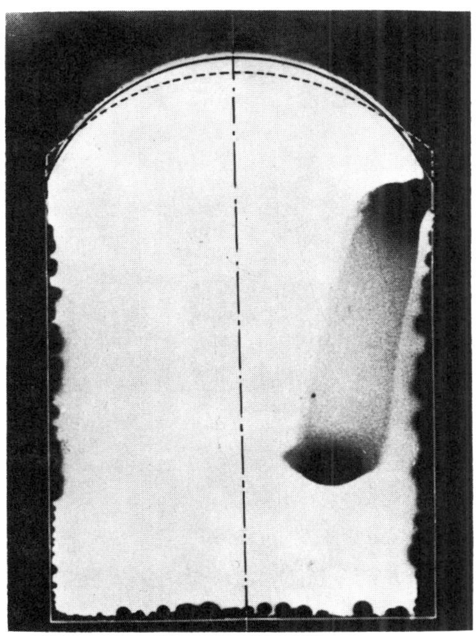

FIG. 6.28. Photograph of a steel melt solidified in an induction furnace and then sectioned. The lines at the top represent the computed meniscus shape.[30] Original photograph kindly provided by Dr. Fautrelle.

parison between the experimentally measured profiles of a laboratory scale molten metal pool and those predicted on the basis of a mathematical model in which allowance is made for the role of the electromagnetic force field in modifying the shape of the free surface.

An alternative measurement technique, sketched in Figure 6.29, relies on the deflection of a rod immersed in a molten metal pool and is applicable to high-temperature systems.[29]

Figure 6.30 shows a comparison between the theoretically predicted and experimentally measured velocities, and their dependence on the coil current, for a 4-ton pilot-scale vessel holding steel. The good agreement is readily apparent.

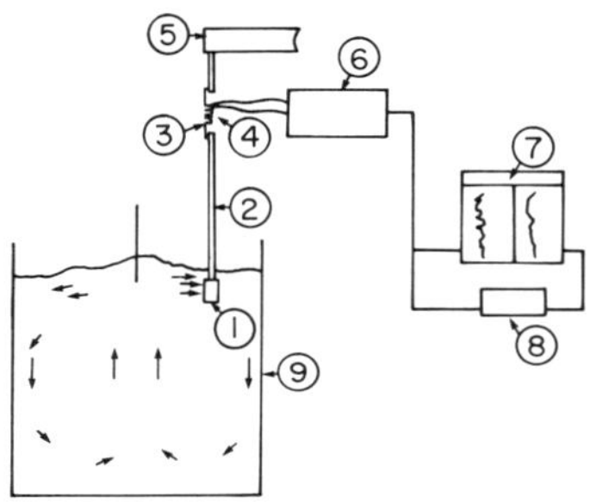

FIG. 6.29. Sketch of the velocity measurement system developed at MEFOS.
1 probe
2 arm
3 spring plate
4 strain gauge
5 support
6 instruments
7 *x*-*x*-*T* recorder
8 rms converter

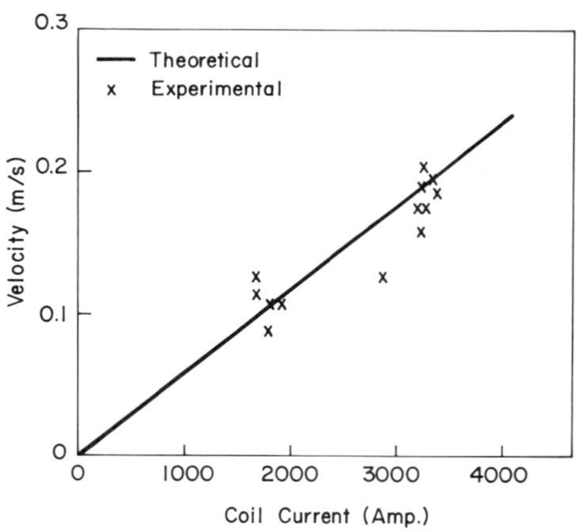

FIG. 6.30. Comparison of the experimentally measured melt velocity (the discrete data points) and theoretical predictions for a 6-ton pilot-scale induction furnace. The plot shows the linear dependence of the velocity on the coil current.

6.3. OTHER LADLE METALLURGY OPERATIONS

The discussion so far has concerned gas bubble–driven circulation systems and induction stirring. For the sake of completeness, we shall present a brief description of two other important ladle metallurgy systems: vacuum degassing units and AOD systems.

6.3.1. Mixing in Vacuum Degassing Systems

While originally intended for vacuum degassing, most of the present-day application of these systems, such as R-H or D-H units, concerns mixing and agitation, since most of these units operate at 100–200 mmHg pressures.

Figure 6.31 shows a schematic sketch of an R-H degassing unit. By the

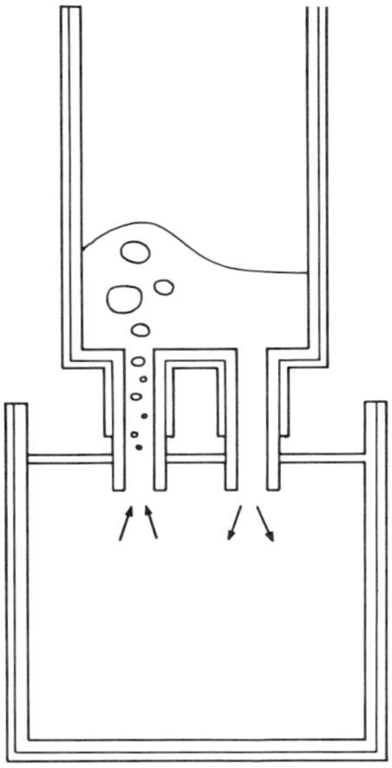

FIG. 6.31. Schematic sketch of an R-H vacuum degassing system.[35] Reproduced with permission of the copyright holder, the Iron and Steel Institute of Japan.

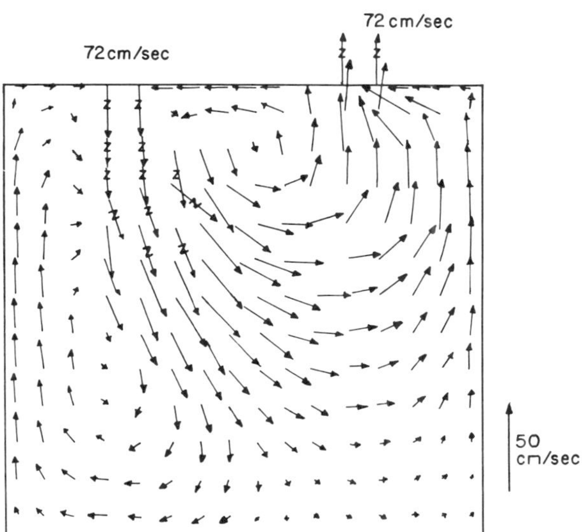

FIG. 6.32. Computed velocity field in an R-H vacuum degassing system, after Shirabe and Szekely.[35] Reproduced with permission of the copyright holder, the Iron and Steel Institute of Japan.

imposition of a vacuum in the upper chamber, a circulatory motion is induced in the ladle. Actual computed velocity fields are given in Figure 6.32.

The velocities in this system are relatively high in the vicinity of the up and the down legs, and relatively low near the bottom of the vessel. The map of the turbulent kinetic energy (not reproduced here) shows a similar pattern. In terms of uniformity in velocity and turbulent energy distribution, these systems appear to occupy an intermediate position between argon-stirred ladles and inductively stirred systems.

6.3.2. The AOD System as a Ladle Metallurgical Tool

Originally intended for the decarburization of stainless steel, the AOD system is gaining application as a ladle metallurgical tool. In an AOD unit, as sketched in Figure 6.33, a gas stream is injected through side-mounted tuyeres at a high velocity, providing excellent agitation. Figures 6.34 and 6.35 show the experimentally measured velocity fields and maps on the turbulent energy obtained in a water model. Both the velocities and the maps of the turbulent kinetic energy are quite uniform, indicating the good mixing attainable in such systems. The absolute values of the turbulent energy levels existing in the real industrial units could not be reproduced in the water model. Thus, the AOD unit provides the highest turbulent energy input in all ladle metallurgy systems.[28]

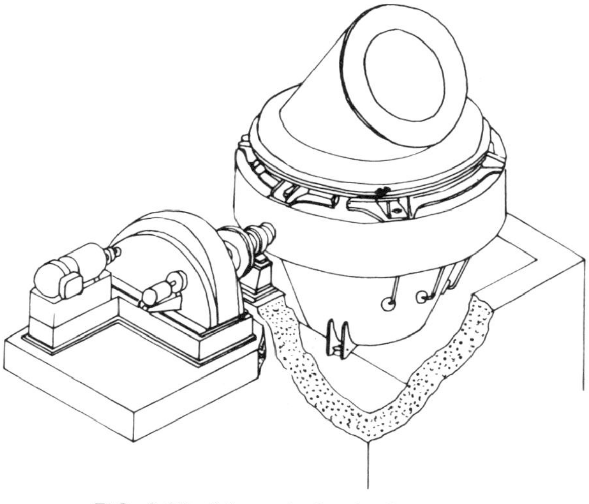

FIG. 6.33. Schematic sketch of an AOD vessel.

6.4. DISCUSSION

The mathematical modeling, physical modeling, and plant-scale experiments discussed here in relation to argon-stirred ladles, induction-stirred vessels, R-H degassers, and AOD units provide an overall picture of the general behavior of

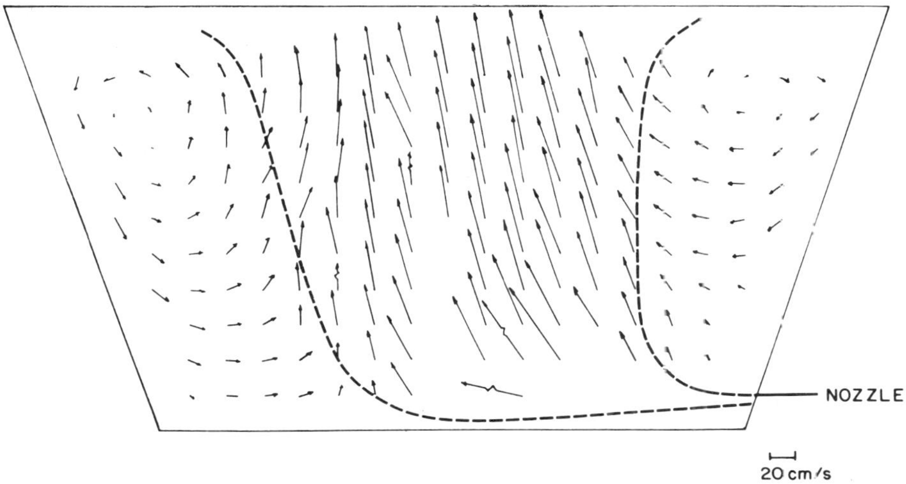

FIG. 6.34. Experimentally measured velocity maps in a water model of an AOD vessel, after Figuera and Szekely.[32]

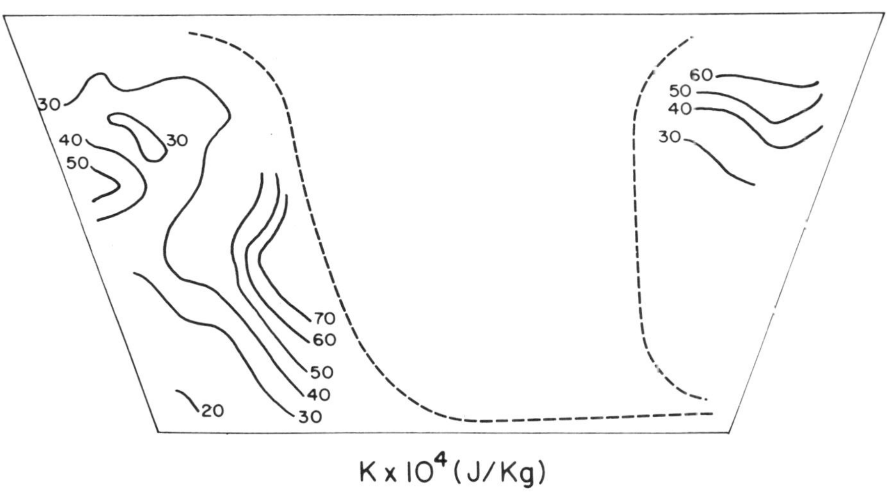

(*a*)

FIG. 6.35. Experimentally measured maps of the turbulent kinetic energy in a water model of an AOD vessel, after Figuera and Szekely.[32]

these systems. An understanding of these systems will greatly aid in process selection.

If the main concern is the extent of agitation or homogenization in a nonreacting system, then Figure 6.23 provides perhaps the best summary of the situation. On plotting the specific energy input per unit mass against the mixing time, one may represent in an approximate fashion all the systems considered. A useful generalization of this plot for a range of geometries has been recently presented by Sano and Mori.[31]

Of the well-studied operations, the inductively stirred vessel with the highest specific energy input provides the shortest mixing time, while the argon-stirred units and R-H degassers may be comparable, depending on the operations parameters involved.

The AOD unit provides by far the highest energy dissipation, and for that reason it should be the best mixer. However, at present there are no direct, industrial-scale measurements to confirm this plausible contention.

The mixing time appears to be quite short for all these systems. Thus, if only homogenization were required, any one of these units could perform satisfactorily.

The problems become subtler when one considers the other functions of ladle metallurgy systems, such as the dissolution of alloying additions, refining (i.e., desulfurization or dephosphorization), inclusion removal, and heating. For these functions, attention must be paid not only to the macroscopic energy input but also to the spatial distribution of the velocities and the turbulent properties of the system.

In comparing the four systems discussed, the induction furnace provides the most uniform distribution, with the R-H and the AOD occupying an intermediate position. In absolute terms, the AOD provides the highest turbulence energy dissipation, with the induction furnace being a close second. Let us consider these problems in detail.

6.4.1. Dispersion of Alloying Additions

Regarding the dispersion of alloying additions, two important steps must be considered: the melting-dissolution kinetics and the subsequent dispersion.

In comparing the possible efficiencies of these systems, it should be noted that in the induction-stirred vessels the velocity fields and the turbulence intensities are high and generally uniform. It follows that rapid dissolution and dispersion occur irrespectively of the location of the introduction of additions. The AOD units should perform even better in this regard. In contrast, in the operation of argon-stirred ladles it is critical that the additions be introduced in the jet plume, where the velocities and the turbulence levels are the highest; this is the practice in the injection processes. If by some operational fault the alloying additions are introduced outside of the plume, the melting and dispersion will be much slower and possibly incomplete.

For R-H degassers the point of introduction is also critical because, unless the alloying additions are introduced in the vicinity of the up-leg or down-leg, the dispersion time can be very slow. This has been proven experimentally.

6.4.2. Reaction Kinetics

In many ladle metallurgical operations the objective is to inject a reactant, such as calcium carbide or magnesium, to remove an undesirable metal component, such as sulfur. For such systems the key features of operation should be (1) to provide high transfer coefficients (i.e., high turbulence levels) in the actual reaction zones and (2) to provide adequate convective mass transfer to convey the reactants to these reaction zones.

In this regard, AOD units and argon-stirred ladles utilizing reactant injection represent the ideal solution, because the turbulent kinetic energy is highest in the jet plume and at the slag–metal interface. where the reactions take place. Furthermore, for the most part the convective mass transfer in the bulk is adequate to supply the reactants to the reaction zone.

In spite of the high velocities and overall high level of turbulent energy dissipation, induction furnaces would not be satisfactory for such operations because of the inadequate agitation provided at the slag–metal interface. Similar considerations apply to R-H degassers.

6.4.3. Coalescence of Inclusions

The coalescence of inclusions, and hence the effectiveness of deoxidation, depends critically on the turbulent energy dissipation in the system. In this regard the induction-stirred systems are likely to be ideal because of the uniformly high velocities and uniform high levels of turbulence energy dissipation. AOD units offer interesting possibilities, but a certain calming period would have to be provided to allow for the flotation of the agglomerated inclusion particles. Argon-stirred ladles are likely to be the least effective because of the existence of quiescent zones at the bottom of the vessel and the large spatial variation in the rates of turbulent energy dissipation. R-H degassers are likely to occupy an intermediate position between induction-stirred vessels and argon-stirred ladles, since the region of high turbulent energy dissipation is in the vicinity of the up leg and down leg, with the remainder of the vessel being less well stirred.

6.4.4. Heating: Ladle Furnace Operation

One important recent development in ladle metallurgy is the more extensive use of ladle furnaces for the addition of superheat. Since R-H degassers and AOD systems cannot function in this manner, the principal question is whether induction stirring or argon stirring would be more effective in the rapid dispersion of the energy supplied by the arc struck between the electrodes to the bath surface.

From the standpoint of temperature control, induction stirring is thought to be preferable by many because of the more regular flow patterns, the more rapid circulation, and the much lower disturbance created at the slag–metal interface. The latter point is of interest because it may allow operation with shorter arcs at a high thermal efficiency.

It should be remarked, however, that another school of thought favors the use of argon stirring and obtaining a high thermal efficiency by burying the electrodes in a foaming slag. This question has not been fully resolved at present.

6.4.5. Containment Wall Erosion

In all agitated systems one has to strive for the maximum allowable energy input to provide the maximum agitation and at the same time to minimize the damage that an intensive circulation can inflict on the ladle lining. Mass transfer is affected by both the melt velocity and the turbulent kinetic energy dissipation in the system. Thus, one would expect the most erosion for the AOD system. Lining erosion is also likely to occur in inductively stirred systems, but this is likely to be uniform along the length of the wall, with more severe erosion in the corner regions. On comparing the anticipated erosion profiles for argon-stirred ladles, induction furnaces, and R-H degassers, one would expect the least erosion in an R-H degasser and perhaps the most erosion at the flux line of an argon-stirred ladle.

6.5. CONCLUDING REMARKS

Through the application of mathematical modeling, physical modeling, and the accumulated plant-scale experience, a reasonably good understanding can be developed of ladle metallurgy operations. All ladle metallurgy operations involve some form of mixing, promoted by gas injection, electromagnetic forces, or vacuum-driven circulation. In addition to mixing, the ladle metallurgy systems function as chemical reactors (desulfurization, dephosphorization, etc.); filters (to remove inclusions); and heaters (ladle furnaces).

The speed with which these operations may be accomplished is markedly affected by the rate at which energy may be transferred into the melt. At the same time, a high level of energy input may also create highly corrosive conditions, making containment more difficult.

In selecting a given system for a particular application, whether it is production of an existing or a new steel grade, careful balance must be struck between cost of installation and performance. Induction stirring, which is likely to be the most expensive, provides a highly uniform velocity and turbulence field, which may be desirable in many applications. However, if slag–metal reactions are to be promoted, gas injection is still necessary. Gas-bubble stirring, through the use of lances or porous plugs, provides a relatively inexpensive means of agitation, although both the velocity fields and the turbulence levels are strongly nonuniform. Vacuum systems, when used as stirrers (such as D-H or R-H) occupy an intermediate position between these extremes. Finally, the AOD system, with the highest specific energy input, appears to offer interesting new possibilities, particularly for situations where a high degree of mixing is desirable.

REFERENCES

1. *Scaninject I: Proceedings of the International Conference on Injection Metallurgy*, MEFOS, Lulea Sweden, 1977.
2. *Scaninject II: Proceedings of the International Conference on Injection Metallurgy*, MEFOS, Lulea Sweden, 1980.

3. J. S. Kirkaldy, ed., McMaster Symposium on Ladle Treatment of Carbon Steel, McMaster University, Hamilton, Canada, 1979.

4. W. G. Wilson and A. McLean, *Desulfurization of Iron and Steel and Sulfide Shape Control*, Iron and Steel Society AIME, Pittsburgh, PA 1980.

5. J. Szekely, H. Y. Wang, and K. M. Kiser, *Met. Trans.*, **7B**, 287 (1976).

6. T. DebRoy, A. K. Majundar, D. B., Spalding, *Appl. Math. Mod.*, **2**, 146–150 (1978).

7. J. W. McKelliget, M. Cross, R. D. Gibson, and J. K. Brimacombe, ICHMT International Seminar on Heat and Mass Transfer in Metallurgical Systems, Dubrovnik, Yugoslavia, 1979.

8. J. Szekely, T. Lehner, and C. W. Chang, *Ironmaking and Steelmaking*, **6**, 285 (1979).

9. H. Tse-Chang, T. Lehner, and B. Kjellberg, *Scand. J. Met.*, **9**, 105 (1980).

10. J. Szekely, N. H. El-Kaddah, and J. H. Grevet, Paper No. 5, in *Scaninject II: Proceedings of the International Conference on Injection Metallurgy*, MEFOS, Lulea, Sweden, 1980, p. 1.

11. J. H. Grevet, J. Szekely, and N. H. El-Kaddah, *Int. J. Heat Mass Transfer* **25**, 487 (1982) and **27**, 1116 (1984).

12. T. Deb-Roy and A. K. Majumdar, *J. Met.*, Nov., 421 (1981).

13. K. Nakanishi, T. Fujii, and J. Szekely, *Ironmaking and Steelmaking*, **2**, 193 (1975).

14. G. B. Wallis, *One-Dimensional Two-Phase Flow*, McGraw-Hill, New York, 1969.

15. B. E. Launder and D. B. Spalding, *Mathematical Models of Turbulence*, Academic Press, New York, 1972.

16. B. E. Launder and D. B. Spalding, *Comput. Meth. Appl. Mech. Eng.*, **3**, 269 (1974).

17. TSI, LDA Instruction Manual, Thermol Systems, Inc., Minneapolis MN.

18. N. H. El-Kaddah and J. Szekely, *Ironmaking and Steelmaking*, **6**, 269 (1981).

19. R. Shinnar and J. M. Church, *Ind. Eng. Chem.*, **52**, 253 (1960).

20. S. Middleman, *AIChE J.*, **11**, 750 (1965).

21. Y. Saro, N. Yamaguchi, and T. Adachi, *J. Chem. Eng. Japan*, 7, 255 (1974).

22. E. Tarapore and J. W. Evans, *Met. Trans.*, **7B**, 343 (1976).

23. J. Szekely, C. W. Chang, and R. E. Ryan., *Met. Trans.*, **8B**, 333 (1977).

24. D. J. Moore and J. C. R. Hunt, *Proceedings of the Third Bar-Sheva Seminar on Liquid Metal Flows and Magnetohydrodynamics*, Bar-Sheva, Israel, AIAA, 1981, p. 84.

25. J. W. Evans and S. D. Lympany, *Met. Trans.*, **14B**, 306 (1983).

26. N. El-Kaddah and J. Szekely, *J. Fluid Mech.*, **133**, 37 (1983).

27. J. Szekely, C. W. Chang, and W. E. Johnson, *Met. Trans.*, **8B**, 514 (1977).

28. E. Tarapore, J. W. Evans, and J. Langfeld, *Met. Trans.*, **8B**, 179 (1977).

29. J. Szekely, N. El-Kaddah, and G. Carlsson, *Met. Trans.*, **15B**, 633 (1985).

30. J. N. Barbier, Y. R. Fautrelle, J. W. Evans, and P. Cremer, *J. Mecanique Theorique et Appliquee*, **1**, 533 (1982).

31. M. Sano and K. Mori, Paper No. 6, in *Scaninject III: Proceedings of the International Conference on Injection Metallurgy*, MEFOS, Lulea, Sweden, 1983.

32. R. Figuera and J. Szekely, *Met. Trans.*, 16B, 67, 1985.

33. B. M. Tracy, Master's Thesis, University of California, Berkeley, 1980.

34. J-H. Grevet, Master's Thesis, Massachusetts Institute of Technology, 198[illegible].

35. K. Shirabe and J. Szekely, *Trans. ISI Japan* **23**, 465 (1983).

7 MODELING OF THE CONTINUOUS CASTING OF STEEL

The continuous casting of steel has emerged as one of the major developments in steel making in this century. Its adoption, replacing traditional ingot casting, for the production of tonnage steel is virtually complete in countries such as Japan and is being pursued vigorously in other parts of the world. The reasons for the phenomenal growth in the implementation of continuous casting are numerous: high productivity, exceptional yield, excellent product quality, and the ability to produce a wide variety of steel grades.

The development of continuous casting and its successful operation in so many steel companies spanning the globe are due in part to the application of mathematical and physical modeling. Mathematical models have been used most successfully in the prediction of heat flow in the solidifying steel strand as part of an overall strategy to control cooling and thereby to ensure quality in the cast product, particularly with respect to freedom from cracks. Similarly, physical models have played an important role in the development of continuous-casting systems and procedures for the production of high-quality steel containing minimum reoxidation and deoxidation products. Thus, physical models have been employed in the design of tundishes (complete with internal devices) and of pouring nozzles and in the control of pouring streams. The objectives in this case have been to minimize the contact of steel with air; to control mixing in the ladle, tundish and mold; and to maximize the float-out of inclusions.

The use of mathematical and physical models as tools in the development of continuous casting is described in this chapter. However, before examining the models a brief introduction to continuous casting is offered for the uninitiated reader. Continuous casting, as the name implies, has the capability of continuously converting molten steel (or any metal) into a solid, semifinished shape, such as a billet (round or square), bloom, or slab. It does this by employing a reciprocating water-cooled copper mold, without a bottom, as the primary heat extraction device (see Figure 7.1). The casting of a heat is commenced by inserting a movable dummy bar into the mold and then partially filling the cavity with molten steel, after which the dummy bar is slowly withdrawn. Lubricant, in the form of oil for billet casting and mold fluxes for larger sections, is continuously applied. A shell of steel solidifies against the walls of the mold and provides a skin encasing the molten core as the partially solidified strand exits the mold. Cooling below the mold and growth of the solid shell are continued by the application of water sprays, often for a considerable fraction of the casting-machine length. Below the spray chamber, cooling of the steel is solely by radiation and natural convection. In the case of large sections, support must be provided to the steel shell below the mold to prevent excessive bulging due to the ferrostatic pressure; the support system, usually rolls, also extracts some heat. Depending on the section size being cast and the withdrawal rate, the liquid pool may extend over a distance of up to 30 m. Downstream of the point of final solidification the strand is cut, usually by torches, into predetermined lengths.

During casting, steel is poured, more or less continuously, into the mold via a tundish from a ladle, as shown in Figure 7.1. The tundish acts as a surge reservoir, as a means for enhancing float-out of inclusions, and as a distributor, since it usually is feeding steel to several strands simultaneously. In the casting of large shapes, such as blooms and slabs, the steel is poured into the mold through

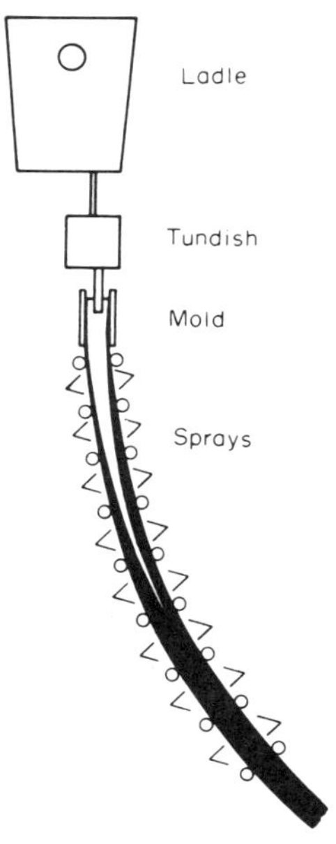

FIG. 7.1. Schematic diagram of continuous-casting machine. Reproduced with the permission of Pergamon Press.

a refractory tube connected to the tundish. The tube is designed to control fluid flow in the mold and to prevent reoxidation of the steel stream by air.

7.1. PHYSICAL MODELS OF THE TUNDISH AND MOLD

The transfer operations and flow phenomena in the ladle, tundish, and mold described above have a significant effect on the quality of continuously cast steel for two reasons. First, during pouring from one vessel to another there is the opportunity, unless preventative measures are taken, for the steel to contact air. The result is reoxidation of the metal with the concomitant formation of large inclusions, which may remain in the steel. Depending on the extent of deoxidation, the steel composition, particularly the deoxidizer content (e.g., Al and Ca), may change significantly in an uncontrolled manner. Second, flow phenomena in the vessels affect both the temperature and the behavior of oxide inclusions originating from both deoxidation, which is usually completed in the ladle, and reoxidation. To control fluid flow in a continuous-casting machine, physical models have been constructed and used to design molds, tundishes, and pouring nozzles as well as to develop operating practices. The principles on which the physical models of tundishes and molds are based have been set out clearly by Heaslip et al.[1]

The flow phenomena to be modeled in the tundish and mold include falling streams (ladle-to-tundish and tundish-to-mold); flow through metering orifices, particularly in the tundish; bulk motion of the liquid steel in both vessels; and possibly wave motion. Moreover, owing to the plunging action of the steel streams, gas is entrained continuously in the liquid and rises to the surface. The flows are all gravity driven, so that, as described in Chapter 3, the Froude number (u^2/gL) is an essential similarity parameter. Thus, any physical model of a tundish and mold must have the same Froude number as the prototype. If the model study is concerned only with flow patterns, such as vortex formation in the tundish or flow through orifices, the Froude number and geometric similarity are adequate scaling parameters. Then, as shown in Chapter 3, from Froude similarity,

$$\frac{u_m}{u_p} = \sqrt{\frac{L_m}{L_p}} = \sqrt{L_s} \tag{1}$$

where L_s is the length scale ratio. For gravity-driven flow through an orifice,

$$u = \sqrt{2gh} \tag{2}$$

where h is the head of liquid above the orifice. The ratio of velocities in model and prototype therefore is

$$\frac{u_m}{u_p} = \sqrt{\frac{h_m}{h_p}} = \sqrt{h_s} \tag{3}$$

Thus, from Eqs. (1) and (3),

$$h_s = L_s \tag{4}$$

and it is seen that the head of liquid in the model tundish must be scaled to that in the prototype according to the length scale ratio.

Similarly, scaling of time can be determined from scale ratios for velocity ($u_s = u_m/u_p$) and length:

$$t_s = \frac{L_s}{u_s} \tag{5}$$

Substituting from Eq. (1),

$$t_s = \sqrt{L_s} \tag{6}$$

Then the scale ratio for flow rate ($Q_s = L_s^3/t_s$) is

$$Q_s = L_s^{5/2} \tag{7}$$

The scale ratio for the orifice diameter can be calculated from Q_s:

$$Q_s = \frac{u_m A_m}{u_p A_p} = \frac{u_m d_m^2}{u_p d_p^2} \tag{8}$$

where A is the cross-sectional area of the orifice. Substituting from Eqs. (1) and (7),

$$d_s = L_s \tag{9}$$

so that the orifice diameter is scaled in conformity with the length scale ratio. From this simple analysis, it is clear that provided only the Froude number and geometric similarity are considered, the model may be scaled to any size relative to the prototype.

Consideration of these two similarity parameters alone is not sufficient if the model is to be used to study mixing in the tundish or mold, however. Then, because the turbulence intensity influences the mixing process in the bulk of the liquid and this is governed by inertial and viscous forces, the Reynolds number assumes importance together with the Froude number and geometric scale ratio. If all three similarity parameters are considered and water is chosen as the model liquid, the model must be made full scale, as indicated in Table 7.1.

Water models also have been constructed of molds based on considerations of Weber and Froude numbers to study the flow pattern produced by different submerged pouring tubes.[2] In this case, as indicated in Chapter 3, the model is attempting to achieve similarity of the surface tension force as well as inertial

TABLE 7.1. Normalized Values of Dimensionless Similarity Parameters for Steel and Water Model

Similarity Parameter	Steel	Water Model
Scale ratio	1	1.0
Froude number	1	1.0
Reynolds number	1	1.1
Weber number	1	3.1
Morton number	1	44.0
Modified Froude number	1	7.0

SOURCE: From Heaslip et al.[1]

and gravitational forces. When the model fluid is water, the scale ratio of such a model must be 0.6.[1]

It is not possible to construct a water model that obeys all four similarity criteria simultaneously. Thus, the study of the two-phase gas–liquid region beneath plunging jets that carry entrained gas into the liquid must be undertaken with considerable caution. The quantity of gas entrained and the size of the gas bubbles are unlikely to be similar in a water model and steel prototype. The rise velocity of the bubbles, which is related to the Morton number ($g\mu^4/\rho\sigma^3$), will not be the same either. As can be seen in Table 7.1, the Morton number in a full-scale water model is 44 times greater than in the prototype. The local flow pattern caused by the gas bubbles, which may be characterized by the modified Froude number [$\rho_g u_g^2/(\rho_l - \rho_g)gL$] will not be simulated faithfully by the water model, since its value is seven times greater than that of the steel system. Nonetheless, there is strong evidence in the literature[1] that, provided the quantity of entrained gas is about the same in model and prototype, the flow patterns in a full-scale water model approximate those in steel reasonably well.

Numerous examples can be cited in which water models have been used successfully to study fluid flow in continuous casting. The influence of the turbulence created in the tundish by the plunging ladle stream on the smoothness of open-pour tundish streams has been investigated by Maddever et al.[3,4] They found that with the ladle stream turned off, the three streams from the tundish under study were smooth, as shown in Figure 7.2*a*. However, when the ladle stream was on and falling between nozzles 1 and 2 (left to right), the streams from these nozzles became very rough, Figure 7.2b. They further showed that the effect of the ladle stream on the roughness of the tundish stream could be minimized by using a castellated nozzle with eight slots cut vertically 16 mm into the top. This was important because stream roughness affects the quantity of air entrained by the tundish stream in the mold pool and hence the extent of reoxidation of the steel. The reoxidation causes large oxide inclusions to form, giving rise to scum on the surface of the pool, which is swept over the

FIG. 7.2. Tundish streams 1, 2, and 3 (left to right) modeled by water (*a*) with ladle stream off (*b*) with ladle stream falling between nozzles 1 and 2.[3] Reproduced with the permission of the Canadian Institute of Mining and Metallurgy.

side to become entrapped as a slag patch in the billet surface. Thus, the water model helped to link a surface quality problem to fluid flow and provided the knowledge to seek a solution with the introduction of the castellated nozzle.

Another example is the study of fluid flow in the tundish of a slab caster conducted by Kemeny et al.[5] Flow was studied by introducing dye into the trough-shaped tundish which fed two slab molds. In the absence of a submerged pouring tube between the ladle and tundish and without flow control devices, such as weirs and dams, in the tundish, several unfavorable conditions were found. Air was entrained in the molten steel in the tundish, and reoxidation resulted. Stagnant regions existed in the tundish pool, thus reducing the effective volume of the vessel. Weirs and dams were introduced, as shown in Figure 7.3, to control flow and eliminate the stagnant zones. Moreover, by adjusting the position and wetted height of these internal devices, the steel was forced to flow in close proximity to the surface slag, which enhanced removal of nonmetallic inclusions. Introduction of a submerged pouring tube between the ladle and tundish effectively eliminated reoxidation of the molten steel during transfer. Again it is seen that the water model is a powerful tool in the achievement of optimum tundish design from the standpoint of maximizing steel cleanness.

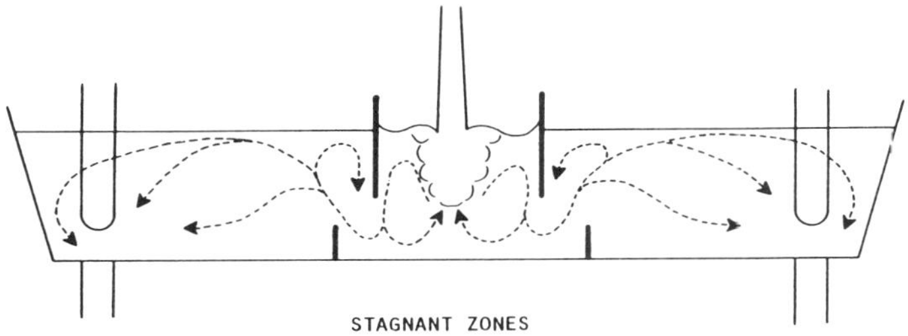

FIG. 7.3. Flow pattern in a tundish equipped with weirs and dams.[5] Reproduced with the permission of the Iron and Steel Society.

7.2. MATHEMATICAL MODEL OF HEAT FLOW IN THE SOLIDIFYING STEEL STRAND

As mentioned at the outset of this chapter, mathematical models of heat flow in the steel have been used as tools to diagnose problems and to control cooling in continuous casting. This is particularly true with regard to the prevention of internal and surface cracks, which are a consequence of thermal and mechanical forces acting on the solidified shell. For example, a heat-flow mathematical model can aid in the determination of the location in the continuous-casting machine at which cracks form. This is because most of the cracks are generated very close to the solidification front, where the steel is exceptionally weak and nonductile owing to the presence of liquid films enriched with segregated elements that separate growing dendrites.[6] Thus, the position of the cracks in a cast section—that is, the distance between the outside tip of the cracks and the surface of the section—effectively gives the shell thickness at the time of crack formation. If a heat-flow mathematical model has been employed to predict the shell thickness as a function of distance from the surface (meniscus) of the molten pool, the shell thickness at which cracks formed can be matched against the predicted shell profile and the location, relative to the meniscus, of crack formation pinpointed. As shall be seen, the mathematical model can also be used to design parts of the casting machine, such as the spray system, to minimize the formation of cracks resulting from thermally generated stresses.[7] Finally, the mathematical model yields much important information, such as the length of the liquid pool in a given or planned casting operation. One would not like to operate or design a casting machine in which the liquid pool extended beyond the cut-off stand. The objectives of a study in which a heat flow mathematical model is to be developed could be addressing any of these aspects of continuous casting. In any case the mathematical model will predict the temperature distribution in the strand as a function of process variables (section size, casting speed, steel composition, pouring temperature, and cooling practice).

Before formulating the mathematical model, it is necessary to review the major heat flow processes operating in continuous casting. Steel entering the

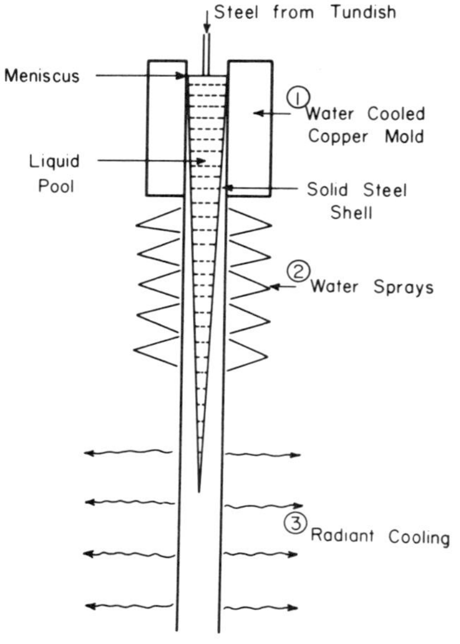

FIG. 7.4. Heat transfer zones in a continuous-casting machine.

mold from the tundish is slightly superheated (10–70°C) relative to the liquidus temperature, to prevent the freeze-up of nozzles and the formation of excessive solidified skulls in the tundish. This superheat is extracted from the molten pool by conduction and convection in the liquid caused by the momentum of the input steel stream. Heat is removed continuously from the surface of the solidified strand as it moves successively through the three heat transfer zones—mold, sprays, and radiation cooling—shown in Figure 7.4. Within the solid shell, heat is conducted from the hot solidification front to the colder surface of the strand; at the solidification front, the latent heat of fusion is extracted. These processes must be incorporated into the mathematical model.

7.2.1. Mathematical Formulation

Once the preparation stage of the mathematical model has been completed (see Chapter 2), the mathematical formulation can commence. A rectangular section having dimensions of X by Y will be considered. The following assumptions are made:

1. The process is operating at steady state, at least with respect to a fixed coordinate framework.
2. A consequence of assumption 1 is that the casting speed, liquid level, and input metal temperature are constant.
3. The rectangular section being cast is symmetrically cooled (i.e., the rate of heat extraction from opposite faces is the same).

4. Heat flow in the liquid pool is by conduction only, but convection is accounted for by artificially raising the thermal conductivity.
5. Steel density is constant.
6. Conduction of heat in the axial (z) direction is negligible.

Next, an infinitesimal volume element, shown in Figure 7.5, is defined and fixed in space such that the strand passes through it at velocity u. Note that in Figure 7.5 the volume element is located within the solid shell, and strictly speaking a second volume element should be positioned within the liquid pool. However, because simple conduction is assumed in the liquid (see assumption 4), the heat flow equation for the solid and liquid are the same. Differences in thermophysical properties and connection of the phases at the solidification front will be handled in the numerical procedure used to solve the heat flow equation. Finally, it is to be noted that in addition to conduction, heat also flows due to the bulk motion of the strand in the axial (z) direction.

Following the steps presented in Chapter 2, a heat balance on the volume element yields the following partial differential equation:

$$\frac{\partial}{\partial x}\left(k\frac{\partial T}{\partial x}\right)+\frac{\partial}{\partial y}\left(k\frac{\partial T}{\partial y}\right)+u\rho C_p\frac{\partial T}{\partial z}=0 \tag{10}$$

From a computational standpoint, it is convenient to change the reference framework to that of an observer who moves with the strand at the casting speed

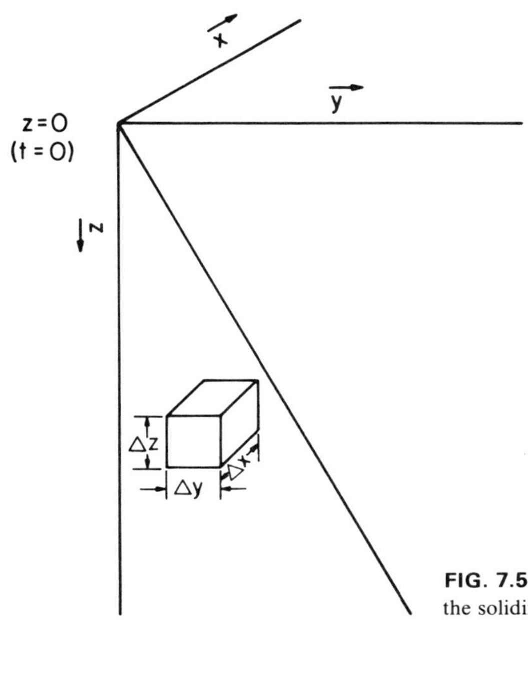

FIG. 7.5. Infinitesimal volume element located in the solidifying strand.

u. Axial position and time, relative to the meniscus, are related simply by the casting speed (i.e., $z = ut$), which may be substituted into Eq. (10). The final equation, then, is

$$\frac{\partial}{\partial x}\left(k\frac{\partial T}{\partial x}\right) + \frac{\partial}{\partial y}\left(k\frac{\partial T}{\partial y}\right) = \rho C_p \frac{\partial T}{\partial t} \tag{11}$$

Thus, this two-dimensional, unsteady-state heat conduction equation must be solved to predict the temperature distribution in a transverse slice of the strand moving through the different heat flow zones of the casting machine. The slice commences its life at the meniscus and does not interact with adjacent slices above or below it because axial heat conduction has been neglected. For the casting of round shapes, the cylindrical coordinate system is applicable and Eq. (11) becomes

$$\frac{\partial}{\partial r}\left(k\frac{\partial T}{\partial r}\right) + \frac{k}{r}\left(\frac{\partial T}{\partial r}\right) = \rho C_p \frac{\partial T}{\partial t} \tag{12}$$

where azimuthal symmetry is assumed ($\partial T/\partial\theta = 0$).

To complete the mathematical formulation, four boundary conditions and an initial condition also must be specified. The initial condition is obtained by specifying the temperature distribution at the meniscus, which is usually taken to be uniform at the pouring temperature, T_p, that is,

$$0 \leqslant x \leqslant X, \qquad 0 \leqslant y \leqslant Y, \qquad T(x, y) = T_p \tag{13}$$

The boundary conditions pertaining to the external surfaces of the strand ($x = 0$, $y = 0$) are derived from a consideration of the continuity of heat flow as follows:

$$x = 0, \qquad -k\frac{\partial T}{\partial x}\bigg|_{x=0} = \dot{q}_0 \tag{14}$$

$$y = 0, \qquad -k\frac{\partial T}{\partial y}\bigg|_{y=0} = \dot{q}_0 \tag{15}$$

where $\dot{q}_0$ is the surface heat flux unique to each of the three cooling zones. The heat flux cannot be determined theoretically owing to the complexity of the surface heat extraction phenomena. Thus, measurements are required of the heat flux profile in the mold as a function of such variables as casting speed, steel grade, and lubricant type.[8,9] Similarly, experiments are necessary to characterize the heat flux in the sprays for different nozzle types; water pressures; nozzle stand-offs; and, where applicable, support roll configuration.[10] This heat flux data, particularly for the spray zone, can be expressed in terms of a heat transfer coefficient. In the radiation cooling zone, $\dot{q}_0$ can be expressed by the Stefan–Boltzmann law:

$$\dot{q}_0 = \sigma \varepsilon (T_0^4 - T_a^4) \tag{16}$$

where the emissivity ε of the oxidized strand surface has a value of about 0.8.

Because symmetrical heat extraction has been assumed, the two remaining boundary conditions can be specified at the center planes ($x = X/2$, $y = Y/2$) of the section where the temperature must be a maximum:

$$x = \frac{X}{2}, \qquad -k\frac{\partial T}{\partial x} = 0 \tag{17}$$

$$y = \frac{Y}{2}, \qquad -k\frac{\partial T}{\partial y} = 0 \tag{18}$$

The unsteady-state heat conduction equation, Eq. (11), and the initial plus boundary conditions, Eqs. (13)–(15) and Eqs. (17) and (18), are a complete mathematical description of the continuous-casting process. The solution of the conduction equation, which yields the temperature distribution throughout the steel strand, can be achieved realistically only by numerical methods. The need to adopt a numerical rather than an analytical approach is a result of the nonlinearity of the boundary conditions and the temperature dependence of the thermal conductivity.[11]

7.2.2. Mathematical Solution

Two different numerical methods have been utilized to solve the conduction equation for continuous casting: finite-difference and integral-profile techniques. The latter, first applied to conduction problems by Goodman,[12] was adapted to the continuous-casting process by Hills.[13,14] It assumes that the temperature profile through the solid shell can be approximated by a truncated power series. The coefficients multiplying each term in the series are evaluated from the boundary conditions at the surface of the solidifying shell and at the solidification front. The power series is substituted into an integrated form of the unsteady-state heat conduction equation, and after considerable mathematical manipulation, two nonlinear differential equations are obtained, which can be solved for the surface temperature and shell thickness as a function of time. This method of solution, although more elegant mathematically than the finite-difference method, has not been accepted widely because in its present form it relies on constant thermophysical properties and does not have the inherent flexibility of the finite-difference method. Nevertheless, the integral profile method has been employed by Brimacombe et al.[15,16] to ascertain the most suitable surface boundary condition for the mold based on overall heat balances on the mold cooling water. Also, it has been used by Miyazawa and Muchi[17] to examine solidification profiles in slab casting. The finite-difference method has been applied to continuous casting by, among others, Mizikar[18] and Brimacombe.[7] For the case of two-dimensional heat conduction, Eq. (11), the alternating-direction implicit method[19-22] is recommended, for reasons given in Chapter 2. Different numerical methods, including the finite-element method, applied to solidification problems have been compared by Thomas et al.[23]

A. Variation of Thermal Conductivity with Temperature

The temperature dependence of the thermal conductivity of steel, k, can be handled in several different ways. One approach, utilized by Mizikar,[18] Kung and Pollock,[24] and others[25,26] involves the approximation of the thermal conductivity by a linear equation:

$$k = a + bT \tag{18}$$

so that the space derivative $\partial/\partial x(k(\partial T/\partial x))$, for example, becomes $k(\partial^2 T/\partial x^2) + b(\partial T/\partial x)^2$. A finite-difference approximation is then substituted for the second term as well as the first. Another method, proposed by Sargant and Slack,[27] relies on the creation of a modified temperature scale ϕ, defined as

$$\phi = \int_{T_d}^{T} \frac{k}{k_d}\, dT \tag{19}$$

which incorporates the effect of varying thermal conductivity. T_d and k_d are reference temperature and corresponding conductivity, respectively. It is also possible that an average thermal conductivity calculated from the temperature of pairs of adjacent nodes (e.g., as between $T_{i,j}$ and $T_{i+1,j}$) could be used in the finite-difference equations.

The thermal conductivity of the liquid steel normally is assigned an arbitrarily high value such as 420 W/mK to account crudely for the convective heat flow as well as conduction. (The convection due to the incoming metal stream is probably confined to the mold and upper spray region.) Fortunately, the precise value of the liquid steel conductivity is not particularly important, since, as mentioned earlier, it affects only the rate at which superheat is removed from the steel; and this is a small fraction of the latent and sensible heats. Further details on the use of the artificially large liquid steel conductivity can be found in Lait et al.[25]

B. Evolution of Latent Heat

The solidification mechanisms that govern the release of latent heat between the liquidus and solidus temperatures are too complex to describe in quantitative terms. However, it has been found that the introduction of complex relationships simulating phenomena such as segregation, dendrite growth, and interdendritic fluid flow are not necessary to preserve reasonable accuracy in the calculation of the temperature field in a strand of steel. Most workers[18,24] have allowed for the evolution of latent heat simply by artificially increasing the specific heat between the liquidus and solidus temperatures. Care has to be exercised in this regard in order to ensure that the exact amount of latent heat is released. Others have found it more convenient to work with enthalpies and to account for the latent heat through the enthalpy–temperature relationship. Further information on this approach is provided in the literature.[7,25]

7.2.3. Model Input and Output

Normally, input data to such a model are

section size,
pouring temperature,
steel composition (which affects the liquidus and solidus temperatures as well as thermophysical properties),
working mold length,
casting speed, and
details of the spray chamber.

The output from the model is the temperature distribution throughout the strand. The solid shell profile can be obtained directly from the temperatures, as it is the locus of elements at the solidus temperature. The usefulness of the results obtained from the model will be discussed in Section 7.2.5.

7.2.4. Comparison of Calculations to Measurements

The best method of verifying the accuracy of the model is to compare model predictions of either shell profile or surface temperatures to measurements made during casting under industrial conditions. The testing of continuous-casting models has been described in several papers.[15,16,25,26] The finite-difference model has been evaluated most extensively and has been found to predict both shell profiles and surface temperatures that agree well with measurements. An example of its accuracy is shown in Figure 7.6. The filled circles on the left-hand side of the plot are values of shell thickness measured from autoradiographs obtained by adding lead plus radioactive gold to the liquid pool of a 14-cm-square billet. The solid line is the shell profile calculated using a finite-difference mathematical model for the same casting conditions. The agreement between the measured and calculated profiles is reasonably good. The thicker shell measured near the bottom of the pool may be due to a lack of penetration of the radioactive gold to the solidification front, defined in the model as the solidus temperature, since near-stagnant conditions probably exist at this depth. On the right hand side of Figure 7.6, agreement between measured and calculated surface temperatures is also good. Thus, on the basis of this and other

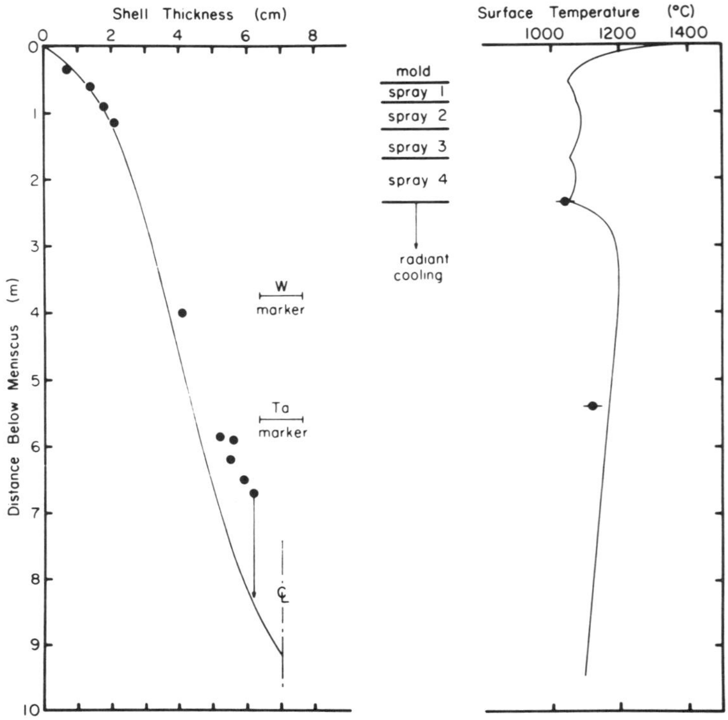

FIG. 7.6. Comparison of model predictions with measurements of shell profile and midface temperature for a 14-cm-square steel billet.[7] Reproduced with the permission of the Canadian Institute of Mining and Metallurgy.

comparisons, it is reasonable to place a good measure of confidence in the ability of the model to simulate the continuous-casting process accurately.

7.2.5. Mathematical Model Calculations

The value of a mathematical model of the continuous-casting process lies in its ability to calculate the temperature distribution in a strand of steel under a wide variety of casting conditions. Its usefulness can be seen in three different situations:

1. Simulation of an existing casting machine with a view to learning more about its operation.
2. Prediction of the effects of a change in a casting variable on the performance of an operating caster.
3. Design of new casting machines.

To those who must run continuous-casting machines on a production basis, the second situation is likely to be of greatest interest. In particular, most process engineers are probably interested in the effect of increasing the casting speed on machine operation as higher throughputs are sought to match planned or existing steel-making capacity. Do changes then need to be made to the mold length, spray system, and position of the cut-off stand? Another area of interest to the process engineer is the minimization of internal defects such as halfway and centerline cracks. How can this be accomplished on a machine that already is in production? These two questions will be discussed in this section as examples of the assistance that a mathematical model, based on heat transfer principles, can give in adjusting casting conditions and improving overall machine performance.

A. Increasing Casting Speed

Increasing the casting speed will have the effect of decreasing the time that the strand spends in the mold and spray zones, and also of increasing the depth of the liquid pool. Looking first at the mold, a decrease in the mold dwell time will result in a thinner shell at the bottom of the mold. Since this may increase the danger of breakouts, an increase in the mold length should be considered. Here the mathematical model can assist us, since it can calculate the shell thickness for different casting speeds and mold lengths. A plot, generated from model predictions, of working mold length (meniscus to mold bottom) versus casting speed is shown in Figure 7.7. The lines correspond to different values of shell thickness at the mold exit. Also plotted in Figure 7.7 are values of working mold length found in industrial practice; the size and shape of the points correspond roughly to the section being cast under those conditions. It can be seen that in

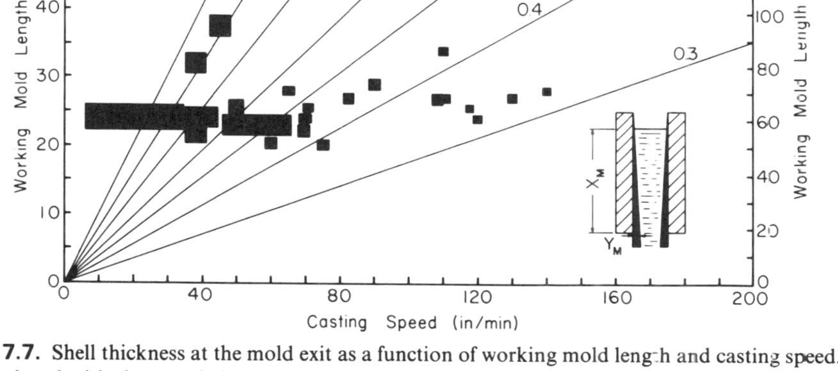

FIG. 7.7. Shell thickness at the mold exit as a function of working mold length and casting speed.[7] Reproduced with the permission of the Canadian Institute of Mining and Metallurgy.

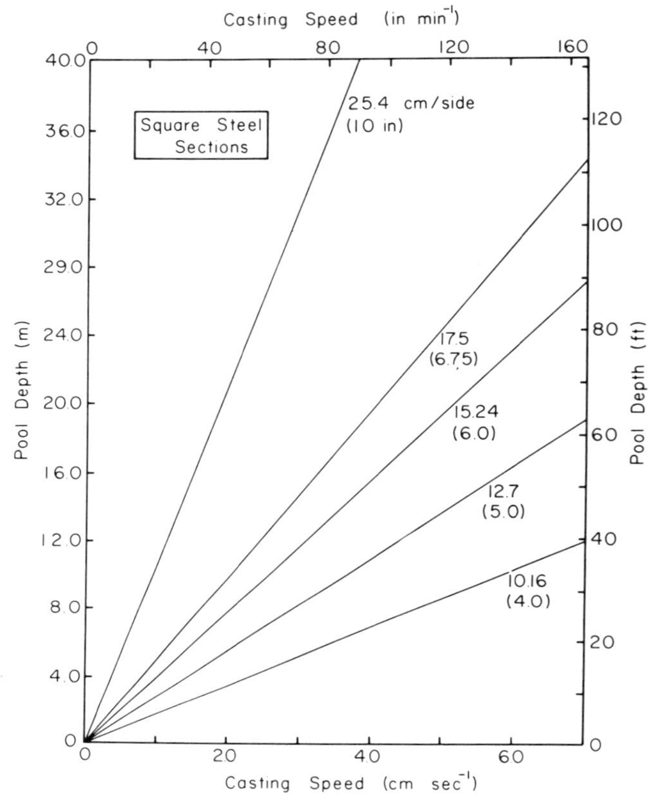

FIG. 7.8. Depth of the liquid pool as a function of casting speed for square billets of varying size.[7] Reproduced with the permission of the Canadian Institute of Mining and Metallurgy.

practice a mold with a constant working length of about 625 mm (25 in.) is used for the casting of the full range of sizes and shapes from slabs to billets. Hence the shell thickness at the mold exit during the casting of slabs is about 20 mm (0.8 in.), whereas for the casting of billets it is roughly 10 mm (0.4 in.). From Figure 7.7 it is clear that if we wish to increase the casting speed, and at the same time maintain a constant shell thickness at the mold bottom, the working mold length should be increased *proportionately*. By this procedure we are simply maintaining a constant mold dwell time, which is the slope of the lines in Figure 7.7.

With regard to changes to the spray system, the length of the spray zones should also be increased in proportion to the increase in casting speed. Again this ensures that the dwell time of the strand in the sprays remains unchanged.

The influence of casting speed on the depth of the liquid pool for several sizes of square billets can be seen in Figure 7.8. Not surprisingly, this shows that the pool depth increases linearly with casting speed. The solidification time, which is the slope of the lines in Figure 7.7, is constant for a given section size; the spray conditions have a relatively small influence on the solidification time. Thus, if the casting speed is to be increased, we must allow for a proportionate increase in the pool depth as well.

B. Altering the Spray System to Minimize Internal Cracks

The changes in a continuous-casting machine that should accompany an increase in casting speed are relatively straightforward compared to the alterations in the spray system that are required to minimize internal cracks such as halfway cracks shown in Figure 7.9, although the principles expounded here apply equally well to minimization of other types of cracks. As will be seen, a mathematical model is an invaluable tool in tackling this problem.

It has been found that halfway, or midway, cracks are the result of reheating of the surface of the strand due to a sudden reduction in the rate of surface heat extraction. This condition may arise between successive spray nozzles that are placed too far apart or between two spray zones, or below the bottom of the last spray zone. Reheating of the surface below the sprays in a real situation can be seen in Figure 7.10. Reheating is the driving force for halfway-crack formation because it causes the surface to expand, relative to the interior and this in turn imposes a tensile strain on the interior of the solid shell near the solidification front where the steel is hot and nonductile.

Clearly if the spray system is to be altered to avoid midway cracks, reheating of the surface of the strand must be minimized as much as practicable. How can

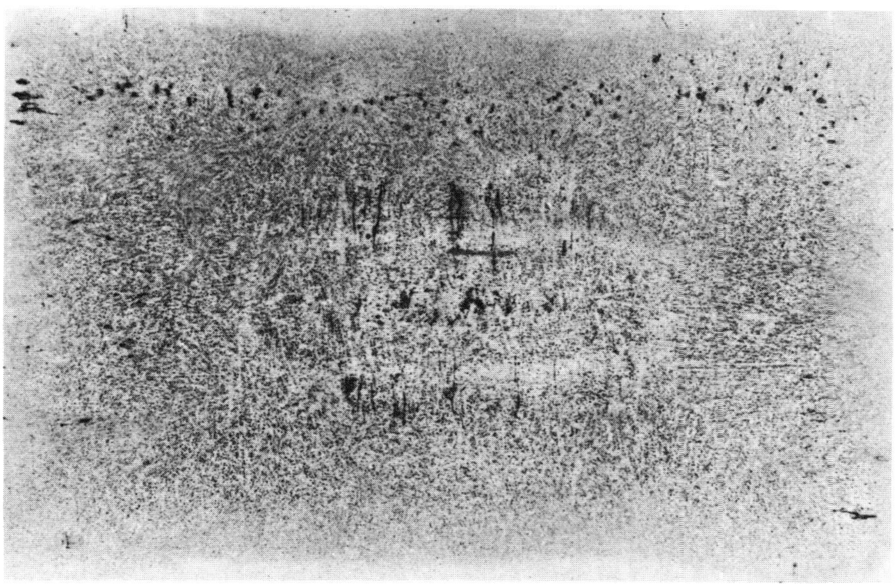

FIG. 7.9. Sulfur print of a transverse section of a billet exhibiting halfway cracks.[7] Reproduced with the permission of the Canadian Institute of Mining and Metallurgy.

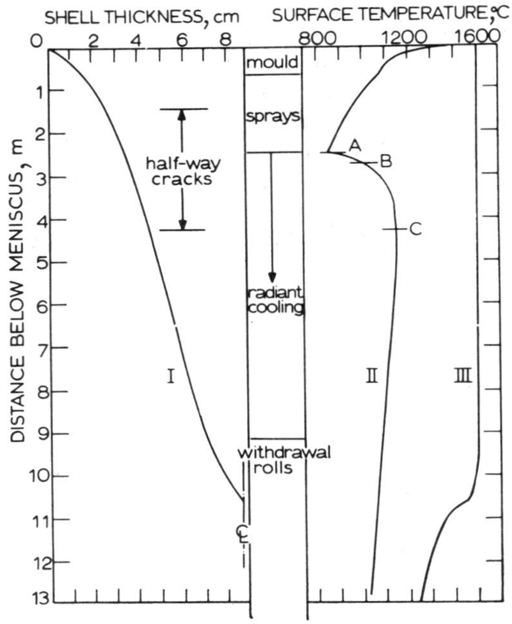

FIG. 7.10. Axial profiles of shell thickness and midface temperature predicted for a 17.1-cm-square billet.[28] Note reheat of the surface below the spray chamber. Reproduced with the permission of the Institute of Metals.

this be achieved in practice (i.e., how can a nonreheating criterion be related to real spray conditions, such as nozzle types, position, and spray pressure)? A mathematical model can provide part of the answer. If the surface temperature distribution through the sprays to minimize subsequent surface reheating is specified, the mathematical model can predict the spray heat flux distribution that is required to achieve it. In the model, the surface temperature profile appears as the surface boundary condition, and from a heat balance on the surface nodes, the surface heat flux distribution can be calculated throughout the sprays. The heat flux distribution then can be converted easily to a heat transfer coefficient distribution, which is a more useful variable. At this point, reliance must be placed on experiments to provide a relationship between spray heat transfer coefficients and practical spray parameters[10] to give the location of spray nozzles and the water pressures to be used. An example of a spray design calculation is illustrated in Figure 7.11 for a 100-mm-square billet. The specified surface temperature at the midface of the billet is shown as the broken line in the center plot. The spray heat transfer coefficient distribution required to obtain the surface temperature is displayed as a contour map in the left-hand plot. The spray heat transfer coefficients can be seen to decrease toward the corners of the billet and with increasing time in the sprays. This observation qualitatively agrees with usual spray practice, where the most intense spraying is given to the

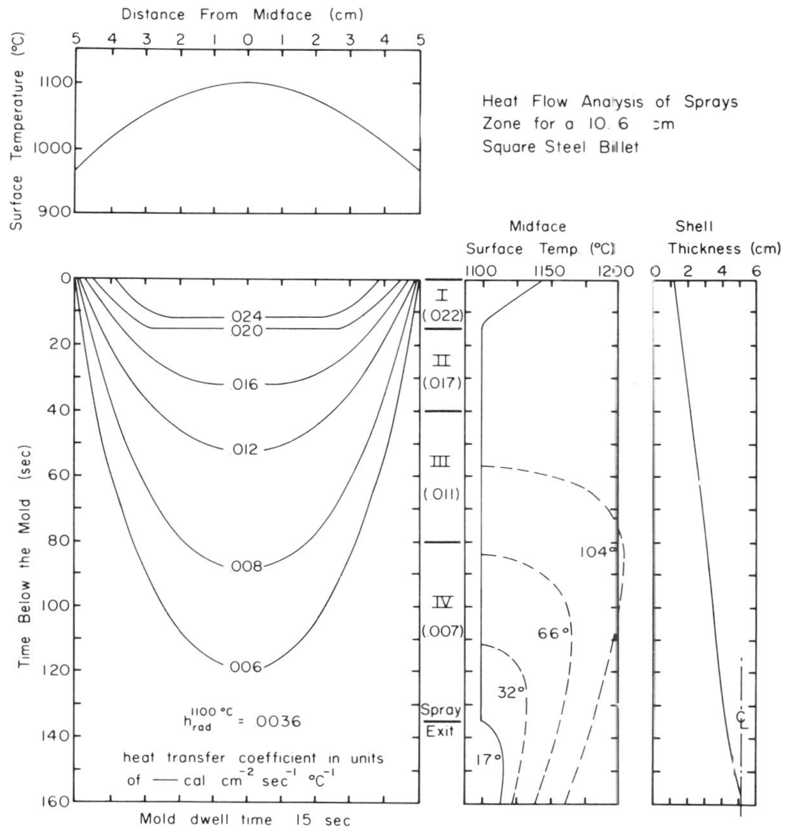

FIG. 7.11. Model predictions of spray heat transfer coefficients, midface temperature, and shell profile for a 10-cm-square billet.[7] Reproduced with the permission of the Canadian Institute of Mining and Metallurgy.

central region of the billet in the uppermost spray zones. Since it is not always practicable to vary the spray heat transfer coefficients continuously, as shown in Figure 7.11, under industrial conditions, the sprays can be divided into zones. Five zones with a constant heat transfer coefficient profile across the face of the billet, have been chosen in this calculation. The profiles are shown in the left-hand plot of Figure 7.12, and the midface temperature profile down the billet can be seen as the solid line in the center plot of Figure 7.11. Spray data reported by Mizikar[30] for a $\frac{1}{4}$ GGIO (Spray Systems) nozzle has been used to relate the spray heat transfer coefficients to spray water flux for each of the spray zones. His data can also be used to relate the water flux to nozzle position and spray water pressure. Thus, by this method using a heat flow mathematical model, it is possible, in principle, to design a spray system to meet any thermal or stress criterion.

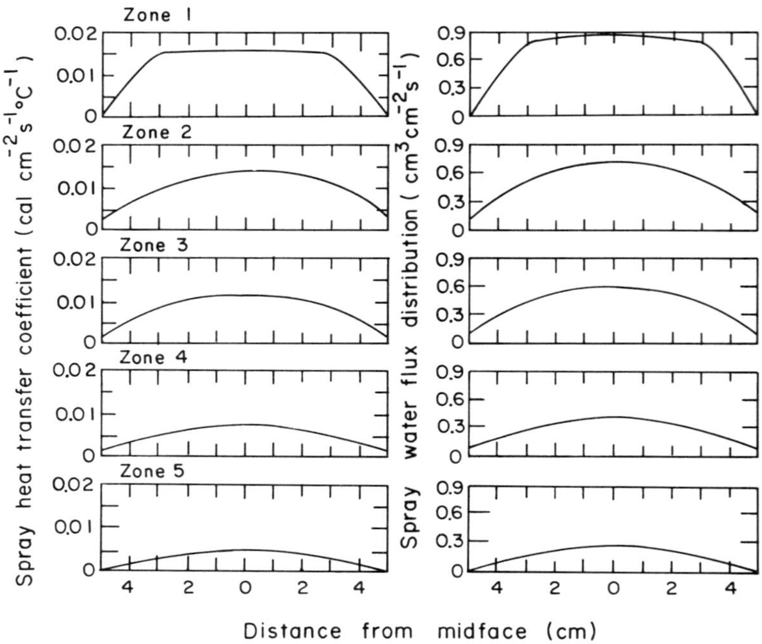

FIG. 7.12. Transverse profiles of spray heat transfer coefficients and spray water flux for a 10-cm-square billet.[29] Five zones are assumed.

7.3. MATHEMATICAL MODEL OF THE MOLD

The mold, which is effectively the heart of the continuous-casting machine, has also been mathematically modeled.[31,32] The model has proven invaluable in elucidating the thermomechanical behaviour of the billet mold and its influence on billet quality (surface and subsurface cracks, deep oscillation marks, and off-squareness or off-roundness).[8,9,33–35] The model also has been utilized to convert measured mold temperatures to axial heat flux profiles for subsequent application in the strand model described in the Section 7.2.1.[8,9] Specification of optimum mold design and operating conditions has been aided by the mathematical model as well.[33–35] It must be emphasized that in each application the model, by itself, would have been insufficient as a tool to reach the project objectives. A series of plant trials and a laboratory study also were required as discussed in Chapter 2. The model and some of its applications are set out briefly in this section.

7.3.1. Mathematical Formulation

The working part of the mold is, for the casting of billets, a water-cooled copper tube and, for larger sections (blooms and slabs) four water-cooled copper plates that are bolted to steel backup plates. The purpose of the mathematical model is to calculate the temperature distribution in the copper mold wall. Ideally the calculation should be three-dimensional in nature, but, owing to a lack of knowledge of steel-to-mold heat flux in the transverse direction, which is required for specification of one of the boundary conditions, heat flow toward the corners of the mold has been neglected. This is not a bad assumption, provided there is no interest in the corner region. Thus, attention is focused on a longitudinal plane, shown in Figure 7.13, cutting axially through the mold wall, and heat conduction in the plane is considered. The following assumptions have been made in formulating the mathematical model:

1. The metal level, or meniscus, is fixed.
2. Mold oscillation is neglected.
3. The water in the cooling channel is in plug flow.
4. Heat transfer from cooling water to the steel liner/backing plate and from the ends of the mold wall is negligible.
5. The cooling channel effectively extends to the ends of the mold.
6. The thermal conductivity of the mold copper is independent of temperature.
7. Axial heat conduction in the water channel is negligible.

From a heat balance on an element of infinitesimal area in the longitudinal plane of Figure 7.13, the following equation is obtained:

$$\frac{k_{\mathrm{m}}}{\rho_{\mathrm{m}} C_{p_m}} \left(\frac{\partial^2 T_{\mathrm{m}}}{\partial x^2} + \frac{\partial^2 T_{\mathrm{m}}}{\partial z^2} \right) = \frac{\partial T_{\mathrm{m}}}{\partial t} \tag{20}$$

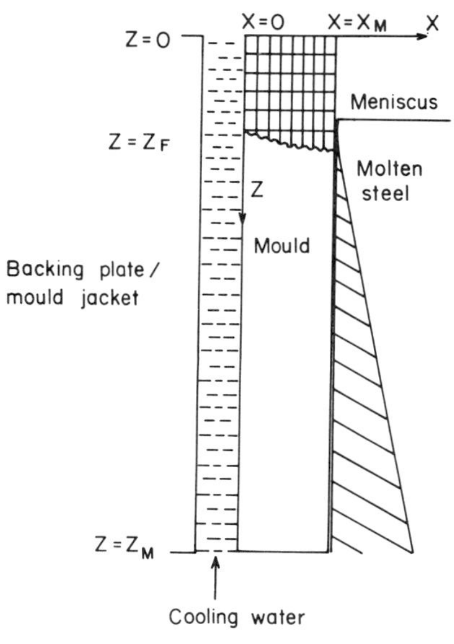

FIG. 7.13. Longitudinal plane through the mold wall.[31] Reproduced with the permission of the Canadian Institute of Mining and Metallurgy.

The time dependence of mold temperature has been retained to account for transient boiling hysteresis. Similarly, a heat balance over an infinitesimal section element in the water channel yields

$$\rho_w V_w d_w C_{p_w} \frac{dT_w}{dz} - h_w(z, t)[T(0, z, t) - T_w(z, t)] = 0 \tag{21}$$

The initial condition for Eq. (20) is

$$t = 0, \qquad 0 \leqslant x \leqslant X_m, \qquad 0 \leqslant z \leqslant Z_m, \qquad T_m = T_m^0 \tag{22}$$

and the boundary conditions are (see Figure 7.13).

Cold face of mold:

$$t > 0, \qquad x = 0, \qquad 0 \leqslant z \leqslant Z_m,$$

$$-k_m \frac{\partial T_m}{\partial x} = h_w(z, t)[T(0, z, t) - T_w(z, t)] \tag{23}$$

Top and bottom of mold:

$$t > 0, \qquad 0 \leqslant x \leqslant X_M, \qquad z = 0, \qquad -k_m \frac{\partial T_m}{\partial z} = 0$$

$$z = Z_m \tag{24}$$

Hot face of mold below meniscus:

$$t > 0, \qquad x = X_m, \qquad Z_F \leqslant z \leqslant Z_m, \qquad -k_m \frac{\partial T_m}{\partial x} = \dot{q}_s(z) \tag{25}$$

Hot face of mold above meniscus:

$$t > 0, \qquad x = X_m, \qquad 0 \leqslant z \leqslant Z_F,$$

$$-k_m \frac{\partial T_m}{\partial x} = h_a(z, t)[T(X_m, z, t) - T_a] \tag{26}$$

Inlet temperature of cooling water:

$$t \geqslant 0, \qquad z = Z_m, \qquad T_w = T_w^0 \tag{27}$$

The steel-to-mold heat flux, $\dot{q}_s(z)$, in Eq. (25) and the heat transfer coefficient at the cooling water–mold interface, $h_w(z, t)$, in Eq. (23) were determined empirically. Values for $\dot{q}_s(z)$ were obtained from measurements on an experimental continuous-casting mold; plots of the axial heat flux profile at the hot face are shown in Figure 7.14 for low- and high-carbon steels. The reduced heat flux in

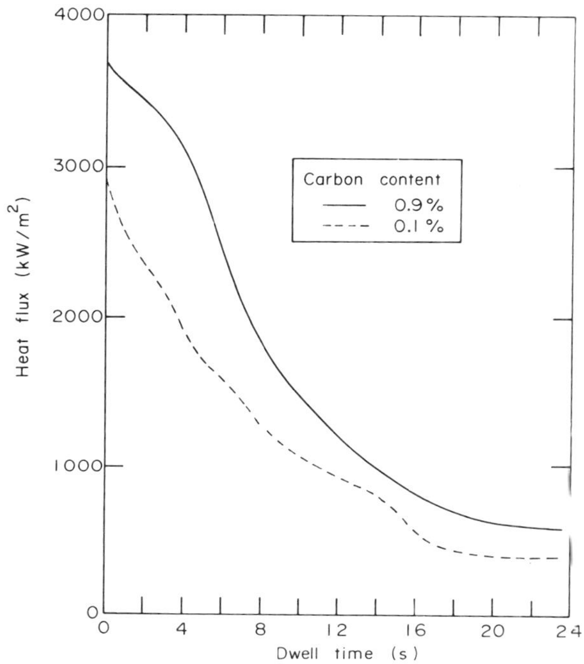

FIG. 7.14. Axial profile of steel-to-mold heat flux at the midface of an experimental continuous-casting mold.[32] Reproduced with the permission of the Institute of Metals.

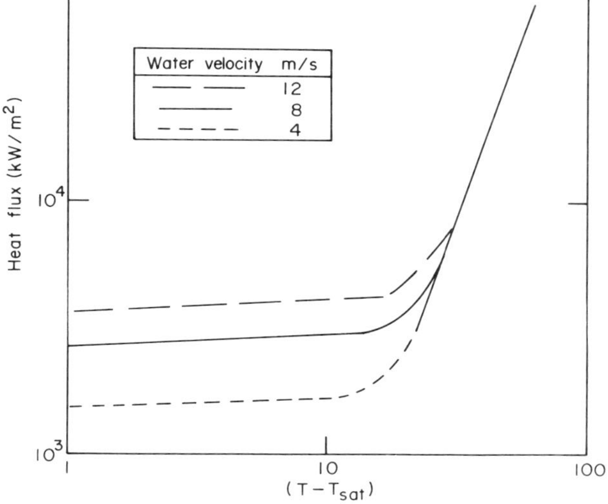

FIG. 7.15. Forced convection–boiling curves for subcooled water flowing at three water velocities.[32] Reproduced with the permission of the Institute of Metals.

TABLE 7.2. Input Data for the Mathematical Model of the Mold

Mold and Casting Variables	Value
Mold length (mm)	700
Wall thickness (mm)	9.53
Section size (mm × mm)	131 × 131
Casting speed (m/min)	2.03
Water inlet temperature (°C)	30
Water velocity (m/s)	7.0
Water pressure in mold (kPag)	241
Mold freeboard (mm)	100
Mold copper alloy	DLP 120 (+Ag)
Direction of water flow	Up
Carbon content of steel	High

SOURCE: From Samarasekera and Brimacombe.[32]

the case of the low-carbon steel is caused by shrinkage due to the δ–γ phase transformation and a concomitant increase in the steel–mold gap width. More will be said about the determination of the heat flux data in Section 7.3.2. Characterization of $h_w(z, t)$ was accomplished with the aid of published correlations which included the influence of boiling.[32] The cold-face heat flux, calculated from the correlations, as a function of local excess mold temperature (relative to the boiling point of the cooling water) is shown in Figure 7.15 for three water velocities.

Equations (20) and (21) were solved using the alternating-direction implicit finite-difference method subject to the initial and boundary conditions. Typical input data to the model for a billet casting mold are given in Table 7.2.

7.3.2. Model Verification

The model has been verified in several ways with measurements made in plant. Axial profiles of mold wall temperature have been measured with the thermocouple arrangement shown in Figure 7.16. The temperatures can be compared to predicted values provided that the steel-to-mold heat flux profile, $\dot{q}_s(z)$ is known. A problem arises, however, because the heat flux depends on numerous variables: lubricant type, casting speed, steel composition, mold distortion, steel superheat, and probably mold oscillation characteristics. Indeed, the heat flux is best characterized using the mathematical model in an iterative fashion, with the heat flux profile being adjusted until predicted and measured temperature match. Samarasekera et al.[9] have taken this approach to generate heat flux profiles in a recent study on a billet casting machine. An example showing the effect of carbon content of the steel is presented in Figure 7.17. Under these conditions the model cannot be verified absolutely, but effectively is tuned to the casting system. However, other measurements—of mold distortion arising from differential thermal expansion—have been made during casting with linear displacement transducers installed in the mold. Measured movement of the mold wall compared reasonably well with predictions based on the heat flow model and a finite-element stress model.[36]

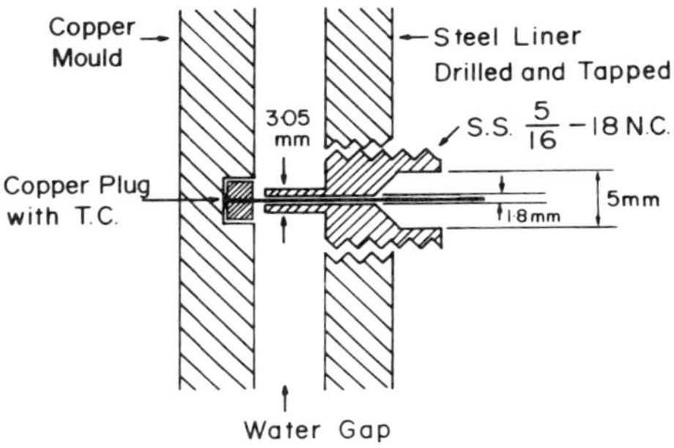

FIG. 7.16. Thermocouple system used to measure mold wall temperatures in a continuous-casting billet mold.[8] Reproduced with the permission of the Iron and Steel Society.

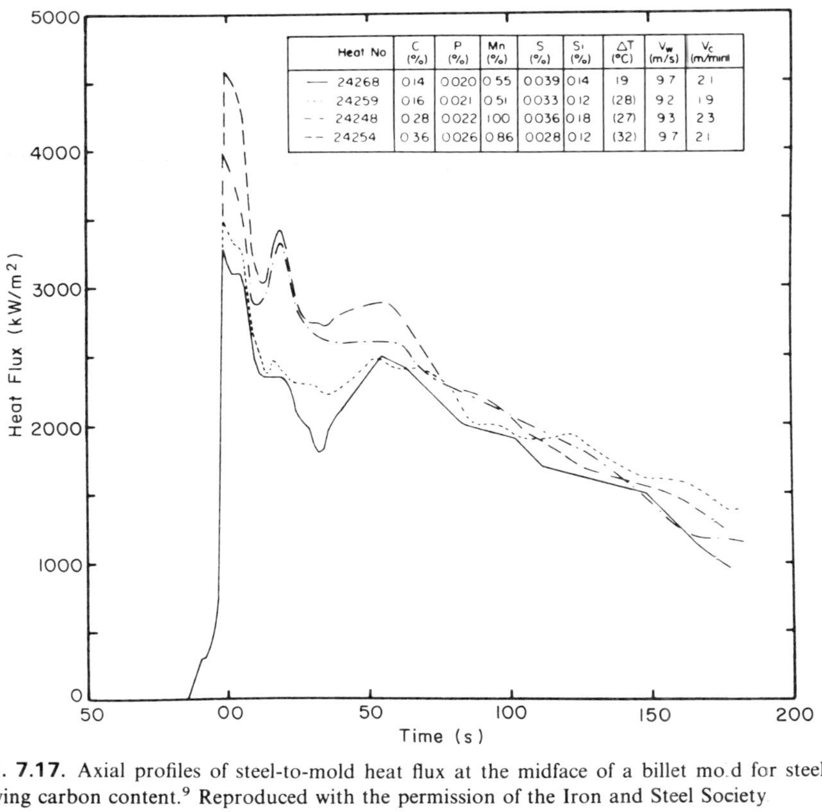

FIG. 7.17. Axial profiles of steel-to-mold heat flux at the midface of a billet mold for steels of varying carbon content.[9] Reproduced with the permission of the Iron and Steel Society.

7.3.3. Model Predictions

The temperature distribution predicted in the mold wall for the conditions given in Table 7.2 is shown in Figure 7.18. An important feature of the predicted thermal field is that the maximum temperature does not reside at the same location as the maximum heat flux, that is, at the meniscus (Figure 7.17), but several centimeters below. The reason, of course, is the strong curvature of the isotherms in the vicinity of the meniscus, which indicates a significant component of conductive heat flux up the mold wall as well as across it to the mold cooling water. This finding indicates that measurement of steel-to-mold heat flux, using pairs of thermocouples spaced a known distance apart, as is often applied in industrial mold trials, will lead to an underestimate of the heat flux. Depending on their relative orientation, thermocouple pairs (combined with Fourier's law) will yield only the through-thickness component of conductive heat flux. The predicted peak temperature below the meniscus also has profound implications, particularly for the continuous casting of billets where the mold consists of a single-piece, unsupported mold tube. Expansion of the tube in the transverse plane is unrestricted, and maximum expansion occurs obviously near

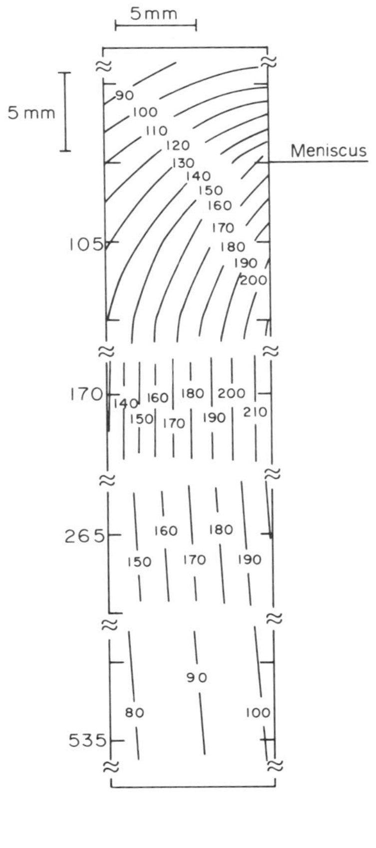

FIG. 7.18. Predicted isotherms (°C) in the wall of a billet mold operating under conditions given in Table 7.2. Reproduced with the permission of the Institute of Metals.

the hottest region: below the meniscus. Thus, during operation, the mold takes on a bulged shape with a negative taper (mold increasing in cross section with axial distance from the top) extending from the top of the mold to below the meniscus. The negative taper, in turn, has an impact on heat extraction and billet quality. It has been shown that conduction across the gap separating the steel from the mold governs mold heat extraction, and therefore the heat flux is dependent inversely on the gap width. Thus, the negative taper adopted by the billet mold has the effect of increasing the gap width and reducing the mold heat flux just below the meniscus; this can be seen in Figure 7.17. Clearly, the thermal efficiency of the mold is reduced.

What is even more important is that the negative taper causes the mold to interact mechanically with the newly forming solid shell at the meniscus during the mold oscillation cycle. Mold reciprocation was introduced originally to continuous casting to prevent sticking between the steel and mold. A critical aspect of the oscillation in this regard is the "negative-strip" period, during which the mold moves downward more rapidly than the descending strand. It can be appreciated that during this time interval, the negatively tapered mold will squeeze down on the solid shell at the meniscus in a jamming action. This

FIG. 7.19. Transverse section of off-square billet.[33] Reproduced with the permission of the Iron and Steel Society.

causes the shell to buckle to form an indentation, or oscillation mark, on the surface of the billet, which again locally reduces heat transfer. If the mechanical interaction between mold and shell is nonuniform around the periphery of the billet, oscillation mark depth and heat extraction also will be nonuniform. The solid shell must necessarily be nonuniform as well, and in the sprays below the mold, differential cooling of adjacent billet faces with varying shell thickness causes the section to be pulled off square (or off round). An example of an off-square (rhomboid) billet can be seen in Figure 7.19. Thus, there is a direct link between temperatures predicted in the mold wall and billet quality.

The nonuniform mold–shell interaction can arise from asymmetrical support of the mold tube[9,33] and from boiling at the cold face of the mold. As seen in Figure 7.18, the maximum cold face temperature approaches 140°C. If nucleate boiling does take place, it may do so intermittently, and differently on adjacent faces. Because boiling locally alters heat extraction by the cooling water, the temperature and expansion of one wall can vary relative to another, giving rise to variable mold–shell interaction. Thus, nucleate boiling at the cold face of the mold is to be avoided to ensure billet quality. The mathematical model can be used to determine the water velocity and pressure that will prevent boiling. The conditions predicted for the casting of medium- and high-carbon steels are shown on a boiling behavior diagram in Figure 7.20.

Finally, the prediction of the mold wall temperature is invaluable for the selection of copper alloys as mold material. The temperature can be compared to the softening characteristics of the alloy to assess its performance and longevity in continuous casting.[31]

Basic knowledge of mold behavior derived from model predictions, combined with industrial measurements and laboratory work, now has led to the specification of an optimum mold design from the standpoint of billet quality.[33–35]

7.4. CONCLUDING REMARKS

Mathematical and physical models have made major contributions to the development of the continuous-casting process. Physical models based on established similarity criteria have been used to simulate fluid flow in the tundish and mold. The objectives have been to control flow to minimize dead zones and short-circuiting of fluid elements so that float-out of nonmetallic inclusions is maximized. The most important similarity parameters are the Froude number, Reynolds number, and geometric scale ratio; and, if water is used as the model fluid, full-scale models are required. The Weber number also has been considered in several modeling studies to achieve similarity of the surface tension force. Examples have been given of physical models being successfully applied to the design of tundish nozzles and the installation of dams and weirs in the tundish.

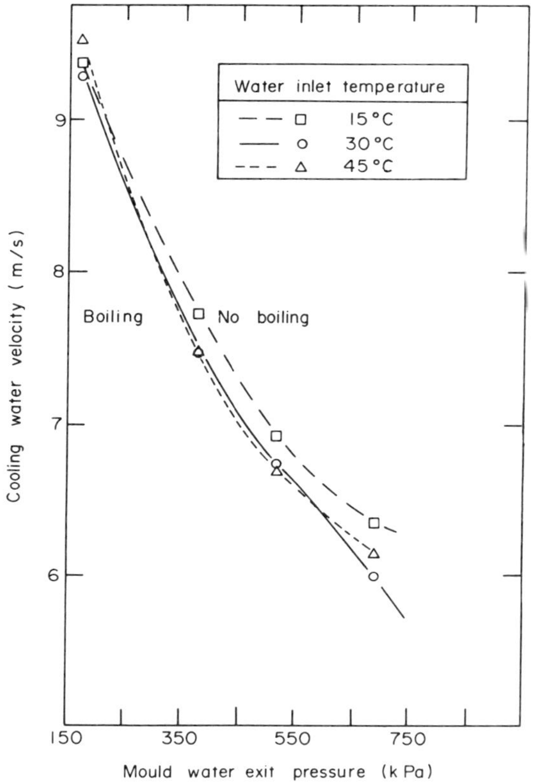

FIG. 7.20. Boiling behavior diagram for the casting of medium- and high-carbon steels (wall thickness, 9.53 mm).[37] Reproduced with the permission of the Metallurgical Society.

Two mathematical models of different regions of the continuous-casting process have been presented. One model has been developed to predict the temperature distribution, as well as the solid shell profile and bottom of the liquid pool in the solidifying strand, as a function of casting variables and design parameters. The model has been validated carefully and subsequently applied to the prediction of the effect of casting speed on the shell profile and length of the liquid pool. Another application is the design of the spray cooling system below the mold to minimize internal crack formation.

The second mathematical model has been developed to predict the temperature distribution in the mold wall of a continuous billet casting machine. The model predictions have not only given rise to a better understanding of heat extraction in the mold but also have established links between its thermomechanical behavior and the formation of defects. This has led to changes in mold design and operation that have improved billet quality.

REFERENCES

1. L. J. Heaslip, A. McLean, and I. D. Sommerville, *Continuous Casting*, vol. 1, Chemical and Physical Interactions during Transfer Operations, Iron and Steel Society, Warrendale, PA, 1983, pp. 67–84.
2. Y. Habu, *Trans. ISI Japan*, **15**, 246–251 (1975).
3. W. J. Maddever, A. McLean, S. Luckett, and G. Forward, *Can. Met. Q.*, **12**, 79–87 (1974).
4. W. J. Maddever, L. Heaslip, A. McLean, J. Beaton, V. Remeika, and L. Hutchinson, AIME Open Hearth Conference Proceedings, vol. 59, TMS-AIME, Warrendale, PA, 1976, pp. 177–211.
5. F. Kemeny, D. J. Harris, A. McLean, T. R. Meadowcroft, and J. D. Young, ISS Second Process Technology Conference on Continuous Casting of Steel, vol. 2, Iron and Steel Society, Warrendale, PA, 1981, pp. 232–245.
6. J. K. Brimacombe and K. Sorimachi, *Met. Trans.*, **8B**, 489–505 (1977).
7. J. K. Brimacombe, *Can. Met. Q.*, **15**, 163–175 (1976).
8. J. K. Brimacombe, I. V. Samarasekera, N. Walker, I. Bakshi, R. Bommaraju, F. Weinberg, and E. B. Hawbolt, *ISS Tran.*, **5**, 71–77 (1984).
9. I. V. Samarasekera, J. K. Brimacombe, and R. Bommaraju, *ISS Trans.*, **5**, 79–94 (1984).
10. J. K. Brimacombe, P. K. Agarwal, S. Hibbins, B. Prabhakar, and L. A. Baptista, *Steelmaking Proc.*, Iron and Steel Society, Warrendale, PA, **63**, 235–252 (1980).
11. British Iron and Steel Research Association, *Physical Constants of Some Commercial Steels at Elevated Temperatures*, Butterworth, London, 1952.
12. T. R. Goodman, *Trans. ASM*, **80**, 335–342 (1958).
13. A. W. D. Hills, J. *ISI*, **203**, 18–26 (1965).
14. A. W. D. Hills, *Trans. TMS-AIME*, **245**, 1471–1479 (1969).
15. J. K. Brimacombe and F. Weinberg, J. *ISI*, **211**, 24–33 (1973).
16. J. K. Brimacombe, J. E. Lait, and F. Weinberg, *Mathematical Process Models in Iron and Steelmaking*, The Metals Society, London, 1975, pp. 174–186.
17. K. Miyazawa and I. Muchi, *Tetsu-to-Hagane*, **60**, 7, 1000–1006 (1974).
18. E. A. Mizikar, *Trans. TMS-AIME*, **239**, 1747–1753 (1967).
19. D. W. Peaceman and H. H. Rachford, Jr., *J. Soc. Ind. Appl. Math.*, **3**, 28–41 (1955).
20. J. Douglas, Jr., *J. Soc. Ind. Appl. Math.*, **3**, 24–65 (1955).
21. J. Douglas, Jr. and H. H. Radford, Jr., *Trans. Am. Math. Soc.*, **82**, 421–439, 1956.
22. B. Carnahan, H. A. Luther, and J. O. Wilkes, *Applied Numerical Methods*, Wiley, New York, 1969.
23. B. G. Thomas, I. V. Samarasekera, and J. K. Brimacombe, *Met. Trans.*, **15B**, 307–318 (1984).
24. E. Y. Kung and J. C. Pollock, *ISA Proc.*, **17**, 8–11–8–12 (1967).
25. J. E. Lait, J. K. Brimacombe, and F. Weinberg, *Ironmaking and Steelmaking*, **1**, 90–97 (1974).
26. J. E. Lait, J. K. Brimacombe, F. Weinberg, and F. C. Muttitt, Fifty-sixth NOH-BOS Conference, Cleveland, 1973, pp. 269–302.
27. R. J. Sargant and M. R. Slack, J. *ISI*, **192**, 428–444 (1954).
28. G. Van Drunen, J. K. Brimacombe, and F. Weinberg, *Ironmaking and Steelmaking*, **1**, 125–133 (1975).
29. A. Grill and J. K. Brimacombe, unpublished research, University of British Columbia, Vancouver, Canada, 1976.
30. E. A. Mizikar, *Iron Steel Eng.*, **47**, 53–60 (1970).
31. I. V. Samarasekera and J. K. Brimacombe, *Can. Met. Q.*, **18**, 251–266 (1979).

32. I. V. Samarasekera and J. K. Brimacombe, *Ironmaking and Steelmaking*, **9**, 1–15 (1982).

33. R. Bommaraju, J. K. Brimacombe, and I. V. Samarasekera, *ISS Trans.*, **5**, 95–105 (1984).

34. I. V. Samarasekera, R. Bommaraju, and J. K. Brimacombe, Proceedings of the International Symposium on Continuous Casting of Billets, CIM Conference of Metallurgists, Vancouver, Canada, 1985.

35. R. J. Dippenaar, I. V. Samarasekera, and J. K. Brimacombe, *Electric Furnace Proceedings*, vol. 43, Iron and Steel Society, Warrendale, PA, 1985.

36. I. V. Samarasekera, D. L. Anderson, and J. K. Brimacombe, *Met. Trans.*, **13B**, 91–104 (1982).

37. I. V. Samarasekera and J. K. Brimacombe, *Met. Trans.*, **13B**, 105–116 (1982).

8 HALL–HÉROULT CELLS

In recent years there has been considerable interest in improving the performance of the electrolytic cells (Hall–Héroult cells) used in the production of aluminum. Such cells consume approximately two-thirds of the total energy involved in the production of aluminum in the United States, or approximately 4% of the electrical energy generated within the nation. Escalating energy costs and occasional shortages of electricity provide great incentives for the improvement of cell performance, especially in the face of increased competition from smelters located in countries with lower energy costs.

Key indicators of cell performance are aluminum production per unit of electric power consumed (kilowatt hours per pound) and aluminum production per unit of current passed (current efficiency). Figure 8.1 shows the historical data on specific energy consumption since early in this century with a projection to the year 2000. Figure 8.2 is a history and forecast of current efficiency (actual aluminum production as a percentage of that expected from Faraday's law). Determination of these indicators requires measurement of mass of aluminum produced, which can be done very precisely (at least over long periods of time), and measurement of voltage and current, which can be done even more precisely. Consequently, improvements of the order of 1% in these indicators can be reliably determined. This, coupled with the incentives referred to above, has led to major efforts by aluminum companies to optimize cell performance. Mathematical and physical modeling have played an important role in this drive toward better performance, and such modeling is being carried out by the major producers of this metal both in North America and overseas.

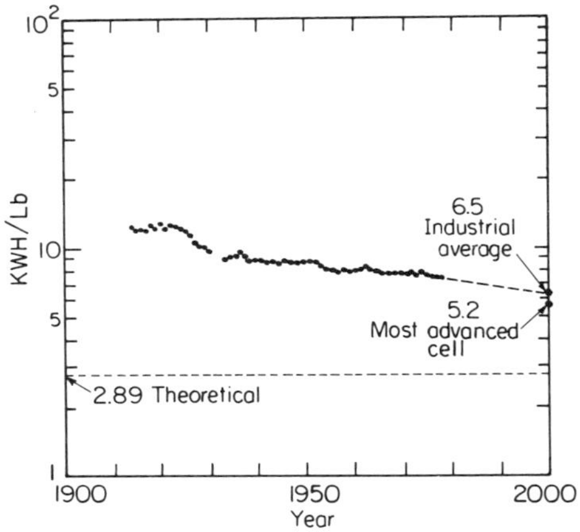

FIG. 8.1. History and forecast of Hall–Héroult cell performance: electrical energy consumption.[1]

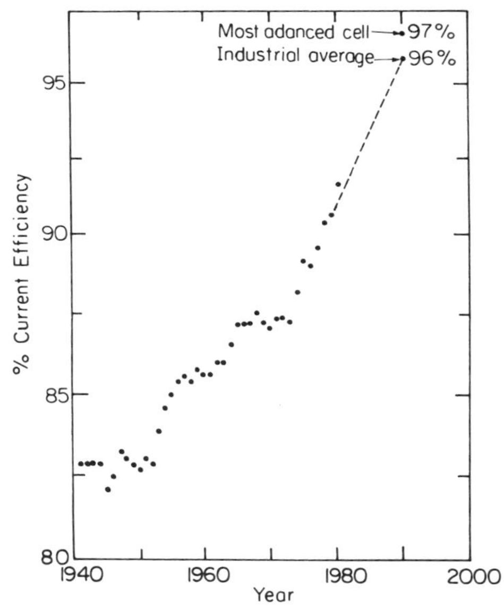

FIG. 8.2. History and forecast of Hall–Héroult cell performance: current efficiency.[1]

8.1. THE HALL–HÉROULT CELL

Figure 8.3 is a cross section of a typical modern cell. The steel shell would be approximately 3 m in the horizontal direction as seen in Figure 8.3; the dimension in the other horizontal direction might be 8 m. Current, of the order of 150 kA, flows to the cell along the anode bus running along the top of the cell. Suspended from the anode bus, by means of anode rods, are carbon anodes that dip into a molten salt electrolyte largely consisting of sodium and aluminum fluorides with dissolved alumina (Al_2O_3) at 950–960°C.

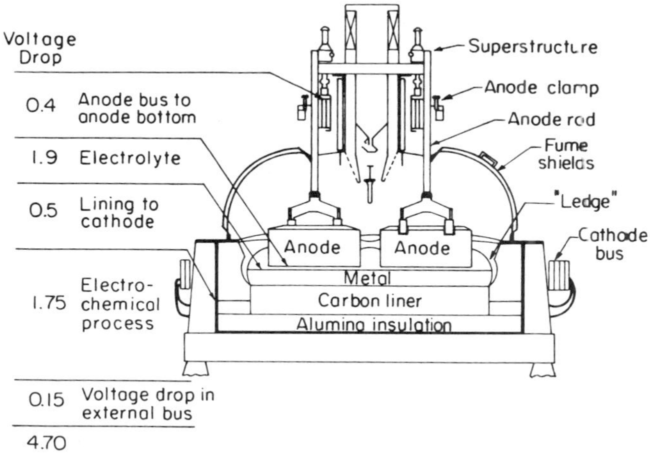

FIG. 8.3. Cross section of a Hall–Héroult cell with approximate voltage drops across various parts (a modern cell would operate at somewhat lower voltage).

This electrolyte is sometimes referred to as *cryolite* (which is its approximate composition) and more frequently as *bath*. Beneath the bath is a pool of molten aluminum, and it is the top surface of this pool that forms the cathodic surface at which the electrolytic reduction of dissolved alumina takes place, forming aluminum. The anodic reaction is the formation of carbon dioxide from the anode carbon. This carbon dioxide passes under the anodes in the form of bubbles, which emerge from the bath alongside the anodes. Beneath the metal pool is a conductive liner of carbon into which steel collector bars project horizontally from the side of the cell. These collector bars are connected to a cathode bus through which the current exits the cell. A typical aluminum plant consists of several hundred such cells. The cells are arranged in lines called *pot lines*, with the cells in each line connected in series. Within each cell electrolysis proceeds with the generation of aluminum and the consumption of carbon anodes. Every 1–2 days, part of the metal is siphoned from the cell, and the anodes (of which there are 10–20 per cell) are replaced in rotation, each one spending 10–20 days in the cell before replacement.

On the left of Figure 8.3 the voltage drops in various parts of the cell are listed. Energy consumption rates are proportional to these voltages, and most cell optimization efforts are aimed at reducing one or more of these numbers. The 1.75 V ascribed to electrochemical processes represents the thermodynamic energy requirements to supply the necessary free energy of reaction. Short of radical changes of cell chemistry (which are outside the scope of this book), little can be done about this value. The other large entry in Figure 8.3 is the voltage drop represented by the resistance of the electrolyte in the interelectrode gap. Many optimization efforts have been aimed at reducing this resistance, frequently by reducing the width of the gap. However, reducing this gap (the anode–cathode distance, or ACD) is no trivial exercise, despite the fact that the anode positions are adjustable by the operator (to compensate for the normal consumption of the bottom of the anode by electrochemical reaction).

A downward adjustment of the ACD in a normally operating cell has two detrimental effects:

1. Because the cell voltage is now less, the rate of electrical generation of heat in the cell is less. Unless other action is taken, the heat balance of the cell is disturbed, and the cell will start to cool. Freezing of the electrolyte occurs at the side of the cell, making cell operation difficult.
2. Short-circuiting between anode and metal is made more probable. Such short circuits cause local overheating, and a fraction of the cell current passes directly from anode to metal without any accompanying electrochemical reaction. Current efficiency and energy consumption per pound of metal produced are then both badly effected. The 1.9 V appearing in Figure 8.3 therefore represent a voltage drop for a typical interelectrode gap of 4–5 cm, which is common practice in industrial cells. This practice is established by cell operators and their foremen, who usually strive to minimize ACD without upsetting the cell by effects (1) and (2). Nowadays such "seat of the pants optimization" is frequently done with the aid of computer control, data acquisition, and mathematical modeling.

Effect (1) can be avoided in design of a new cell (e.g., by installing more insulation to achieve a heat balance at lower ACD), but overcomming effect (2) represents a much more substantial task. The task is made more difficult by uncertainty concerning ACDs within a cell. The bottoms of the anodes are frequently consumed unevenly, and the metal–bath interface is usually deformed by electromagnetic forces (as will be discussed in Section 8.4). Occasionally oscillation of the metal–bath interface occurs. Except for the development of an inert anode, little can be done about the unevenness of the bottoms of anodes; but there is some prospect of minimizing the deformation (and oscillation) of the metal–bath interface, and modeling is playing a role in efforts to do so.

So far, this discussion has concerned the reduction of cell voltage. An alternative approach to improving cell performance is to increase the quantity of aluminum produced per quantity of charge passed, that is, to increase the current efficiency. As can be seen in Figure 8.2, the amount of aluminum produced in a typical cell is approximately 90% of that predicted from theory (Faraday's law). There are many smaller causes of this discrepancy (e.g., "parasitic" currents that flow from anode bus to cathode bus without passing through the electrolyte), but the major cause is thought to be a chemical reoxidation of part of the aluminum product.[2] One possible mechanism by which this could occur is the dissolution of aluminum metal into the bath (measured solubilities have been in the range 0.04%–0.15 w%) followed by mass transport of the aluminum to the vicinity of the anode, where reoxidation by CO_2 bubbles occurs. Mathematical models and measurements on actual cells[3] have shown that the bath is in motion with velocities of the order of 10 cm/s under a combination of electromagnetic forces and bubble stirring. Flow of the bath therefore has an effect on mass transport and thereby on current efficiency.

8.2. MATHEMATICAL MODELS OF THE HALL–HÉROULT CELL: AN OVERVIEW

The Hall cell is a complex unit operation (compared, say, to the relative simplicity of an induction furnace) wherein heat transport, fluid flow, mass transport, bubble effects, chemical and electrochemical reactions, charge transport (electric current), and electromagnetic effects are all important. Furthermore, the majority of these phenomena interact, which makes it inadvisable to consider them in isolation.

For example, Figure 8.3 shows the existence of a ledge (sometimes referred to as *freeze* or *ridge*) at the sides of the cell between the melts and the carbon lining. This ledge of frozen electrolyte between the bath and the carbon serves to protect the carbon from exposure to the bath. Furthermore, it provides a considerable amount of insulation at the sides of the cell; temperature drops across the ledge are several hundred degrees Celsius. Maintenance of a suitable ledge thickness is of great importance in cell operation. Prediction of the ledge thickness has been an objective of thermal mathematical models that have emphasized heat transport in the ledge, walls, and bottom of the cell. Maintenance (or change) of ledge thickness depends on balance (or imbalance) of heat fluxes at the ledge–bath or ledge–metal interface. At steady state, heat convection to this interface from the melt is balanced by heat conduction away from this interface into the ledge. The former is affected by flow of the melt adjacent to the ledge as well as by the melt temperature. Consequently, a rigorous model of heat transport in the cell would have to take into account fluid flow as well as heat generation by dissipation of electrical energy (particularly in the bath).

Conversely, the flow of bath and metal are in large measure due to electromagnetic forces established by the interaction of currents within these melts and the cell magnetic field. The currents within these melts are dependent on the ledge profile and on the (temperature-dependent) electrical conductivity of anodes, carbon lining, and collector bars. It follows that a model for fluid flow in the cell that does not embrace heat transport is incomplete.

No complete model (mathematical or physical) of the aluminum cell has been described in the literature. The published models have included only some of the important phenomena. Despite the incompleteness of these models, they have provided valuable insight into the behavior of real cells; furthermore, they are valuable building blocks from which a complete model will eventually be constructed.

Most mathematical models for Hall–Héroult cells can be categorized as follows:

1. "BLACK-BOX" MODELS, WHICH INCORPORATE EMPIRICAL DATA ON CELL PERFORMANCE AS A FUNCTION OF CELL PARAMETERS, PERHAPS TOGETHER WITH OPERATOR EXPERIENCE AND SIMPLE (OVERALL) HEAT AND MATERIAL BALANCES. Such a model can be useful for predicting the effects of design or operating changes that are within the limits of previous experience. Models of this type typically entail few computations and are therefore valuable when computational costs are significant, for example, for dynamic control of cells when prediction of cell performance (under conditions little different from those currently existing) must be made rapidly and repeatedly. At least one aluminum company uses a model of this kind for training cell operators. Asbjornsen et al.[4] have described a mathematical model of this type for voltage and current fluctuations in a cell. The model carries out a regression analysis of cell voltage and current to estimate, on line, the decomposition potential, bath resistance, and anode position.

2. MODELS FOR CELL VOLTAGE AND CURRENT DISTRIBUTION. Within any of the materials of a cell, the current density is conserved, that is, there are no sources or sinks of current. In mathematical terms,

$$\nabla \cdot \mathbf{J} = 0 \tag{1}$$

that is, the divergence of the current density vector is zero. Furthermore, the current density $\mathbf{J}$ is related to the voltage gradient by the differential form of Ohm's law:

$$\mathbf{J} = -\sigma \nabla \phi \tag{2}$$

where ϕ is voltage, and
σ the electrical conductivity.

As already mentioned, σ is a function of temperature and can therefore be expected to vary with position in the material. A simplification frequently employed in the models under discussion is to presume that the conductivity is constant and at some value corresponding to the anticipated temperature of the material. With a constant conductivity Eqs. (1) and (2) then yield

$$\nabla^2 \phi = 0 \tag{3}$$

This is Laplace's equation, for which a number of analytical solutions or approximations are available, as well as a considerable amount of software. It is usual to assume that ϕ is continuous across the interfaces between materials in the cell (e.g., between the metal and carbon lining). To do so ignores the potential differences due to electrochemical reaction at the interfaces (anode–

bath and bath–metal) where these reactions are occurring. However, these differences in potential are only weakly dependent on current density. Consequently, the potential differences (1.75 V in Figure 8.3) can be left out of the solution of Eq. (3) and then added in afterwards to yield the total voltage for the cell. Knowing the distribution of ϕ permits the calculation of the distribution of current density from Eq. (2) by numerical differentiation.

An example of this kind of model is that of Furman,[5] who was concerned with horizontal current components in the metal pool. As discussed in more detail in Section 8.4, horizontal currents are thought to be detrimental to cell performance. Furman calculated the influence of metal pool depth, anode arrangement, carbon lining, collector bar design, and ledge position on these currents.

3. THERMAL MODELS. These range in complexity from the overall heat balance carried out by Ohta and Matsushima[6] for cells with self-baking (Soderberg) anodes to models where the temperature distribution throughout the cell is computed. A model of the latter variety is discussed in detail in Section 8.4.

4. MAGNETIC MODELS. Once the current distribution throughout the cell has been obtained (e.g., as discussed under category 2), then the Biot–Savart law can be used to compute the magnetic field at any point inside or outside the cell:

$$\mathbf{B} = \mu \iiint \mathbf{J} \times \frac{(\mathbf{r} - \mathbf{r}')}{|\mathbf{r} - \mathbf{r}'|^3} \, dx' \, dy' \, dz' \tag{4}$$

where $\mathbf{B}$ is the magnetic induction at the point of interest (the point represented by the vector $\mathbf{r}$ joining it to the origin of some coordinate system). The integration is carried out over all points (represented by the vector $\mathbf{r}'$ joining them to the origin) at which the current density $\mathbf{J}$ is significant. Hence, currents in external busbars are included in the calculation, as are currents in nearby cells. A thorough discussion of this topic has been provided by Solinas.[7] Important contributions to this subject have been provided by Robl,[8] who has examined the difficulties of using Eq. (4), which is strictly correct only in the absence of ferromagnetic materials, to describe the magnetic field inside an aluminum cell (which has a steel shell).

5. MODELS FOR FLOW OF THE MELTS, FOR INTERFACE DISTORTION, AND FOR MASS TRANSPORT. These models have been concerned with the flow of the bath and metal driven by electromagnetic forces, with the distortion of the bath–metal interface produced by these forces, and with the transport of dissolved aluminum (i.e., current efficiency). Recent models of this kind are those of Tarapore,[9,10] Evans et al.,[11] Arita and Ikeuchi,[12] Lympany and Evans,[13] Lympany et al.,[14] Robl,[15] and Moreau and Evans.[16] A detailed presentation of one of these models appears in Section 8.4.

8.3. THERMAL MODELS FOR HALL–HÉROULT CELLS

Sulmont and Hudault[17] have described a thermal model for a Hall–Héroult cell that was developed by Aluminium Pechiney. The model treats conduction of heat within the sides and bottom of the cell as a two-dimensional heat flow problem (i.e., conduction in the long horizontal direction is treated as negligible). The relevant heat transport equation is then (for steady-state conditions)

$$\frac{\partial}{\partial y}\left(k\frac{\partial T}{\partial y}\right)+\frac{\partial}{\partial z}\left(k\frac{\partial T}{\partial z}\right)+\dot{q}=0 \qquad (5)$$

where y and z are coordinates in the short horizontal direction and vertical direction, respectively,

T is temperature,

k is the (position- and temperature-dependent) thermal conductivity, and

$\dot{q}$ is the electrical heat generation rate per unit volume, obtained from

$$\dot{q}=\frac{1}{\sigma}\,|\mathbf{J}|^2 \qquad (6)$$

We see then that solution of Eq. (5) requires a knowledge of the current distribution within the cell. This is obtained from a solution of Eqs. (1) and (2) with a position- and temperature-dependent electrical conductivity. The temperature dependence of the electrical conductivity mandates that Eqs. (1) and (2) be solved simultaneously with Eqs. (5) and (6). Sulmont and Hudault did this by an iterative numerical solution. Iterative solution is necessary for another reason: the equilibrium position of the ledge is unknown at the start of the calculation and is dependent on the balance between heat input into the ledge surface from the bath or metal and heat conduction into the ledge from this surface. During the iterative calculations the position of the ledge surface is adjusted until this balance is achieved.

Within the bath and metal, convective transport of heat occurs, in addition to conduction, since these melts are in motion. Nevertheless, Sulmont and Hudault used Eq. (5), which ignores convection, for these melts and made allowance for convection by artificially increasing the thermal conductivities. Boundary conditions for Eq. (4) are provided in terms of radiative and convective losses from the steel shell and from the alumina piled on top of the frozen crust of the electrolyte and on anode tops into the cell environment. At interfaces within the cell continuity of fluxes is assumed. For solid–solid interfaces this presents no difficulty, since the fluxes are given by temperature gradients multiplied by a conductivities. For solid–liquid interfaces (e.g., the ledge–metal interface) it is necessary to invoke a heat transfer coefficient. Because this coefficient is difficult to determine from first principles, it is usual to leave it as an adjustable parameter to be fixed by matching the model predictions to measurements on real cells (although Sulmont and Hudault provide no description of how they obtained their coefficients).

Sulmont and Hudault solved the equation using a finite element technique employing 1683 nodes in the example cited in their paper. Following solution of the two-dimensional heat and electrical conduction equations, an allowance was made for heat losses from the end of the cell to arrive at an overall computed heat balance.

The model yields a number of results of practical interest; Figure 8.4 shows the ledge shape and position computed by this model for two Pechiney cells. Isotherms for a 130-kA cell are plotted in Figure 8.5; only half the cell is shown, since the model assumes symmetry about the center plane. The computed horizontal current density in the metal is shown in Figure 8.6. Table 8.1 provides a comparison of the computed results for three types of cells with actual measurements of voltage on these cells; clearly the model is quite successful.

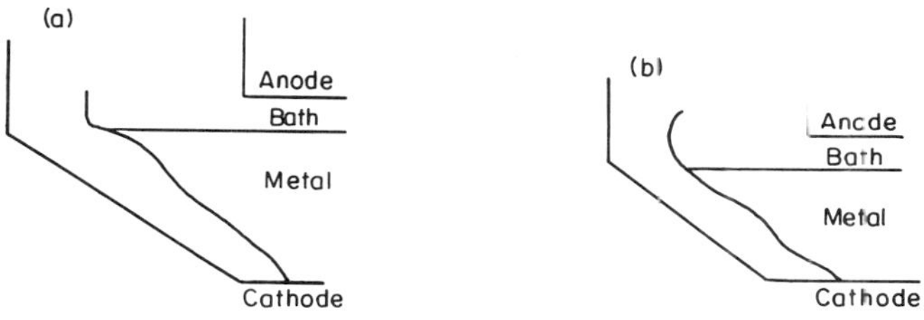

FIG. 8.4. Example of ledge equilibrium profiles computed by a thermal model for (*a*) a 130-kA cell and (*b*) a 90-kA cell. Reprinted with permission from *Light Metals 1978*, John J. Miller, ed., The Metallurgical Society of AIME, Warrendale, PA, 1978.

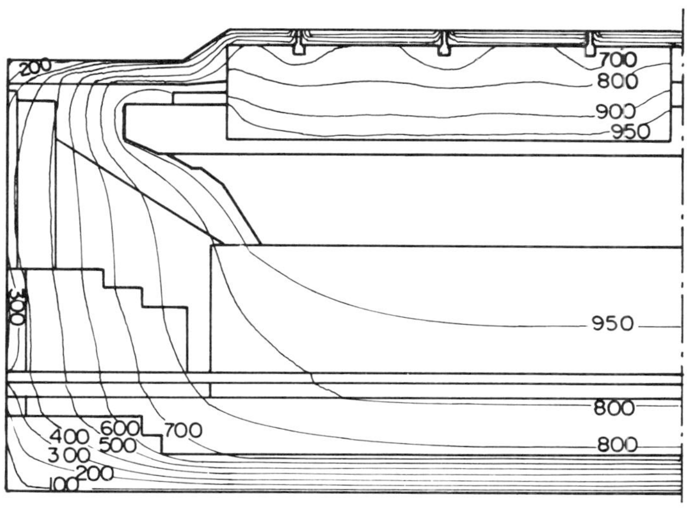

FIG. 8.5. Isotherms computed for a 130-kA cell. Reprinted with permission from *Light Metals 1978*, John J. Miller, ed., The Metallurgical Society of AIME, Warrendale, PA, 1978.

TABLE 8.1. Comparison of Calculated and Observed Results for Three Types of Pot

130-kA Pot, Voltage

	Model	Pot
Anode	0.19	0.19
Cathode	0.37	0.43
Bath + electrochemical voltage	3.14	3.02
Buss loss	0.31	0.31
Total	4.01	3.95

130-kA Pot, Thermal Energy

	Model	Pot
	(kW)	(%)
Sides of pot shell	70	25
Cathode bars	20	7
Bottom of pot shell	25	9
Surface of anodes	85	31
Pins	51	18
Alumina in annular channel between anodes and pot shell	28	10
Total	279	100

72-kA and 145-kA Pots

	72 kA		145 kA	
	Model	Pot	Model	Pot
Anode	0.21	0.18	0.26	0.25
Cathode	0.32	0.36	0.37	0.39
Bath + electrochemical voltage	3.17	3.14	3.15	3.17
Alumina in annular channel between anodes and pot shell	0.22	0.22	0.29	0.29
Total	3.92	3.90	4.07	4.10

SOURCE: From the study of Sulmont and Hudault.[17]

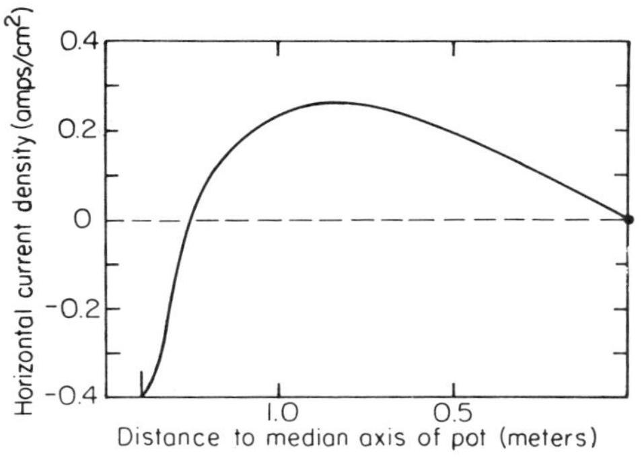

FIG. 8.6. Distribution of the computed horizontal current density in the metal of a Hall–Héroult cell. Reprinted with permission from *Light Metals 1978*, John J. Miller, ed., The Metallurgical Society of AIME, Warrendale, PA, 1978.

Sulmond and Hudault report on the use of the model to predict the effect of changes in operating or design parameters. For example, Figure 8.7*a* shows the effect of a 2% increase in cell current on the shape and position of the ledge. Figures 8.7*b* and 8.7*c* contrast the ledge location at two different metal depths. The shallower metal pool results in higher current densities in the pool, which result in greater electromagnetic stirring forces and thereby in a higher heat transfer coefficient between metal and ledge. Reducing the depth of the metal pool would therefore cause the ledge to melt back to a new equilibrium position where the conductive heat flux into the ledge again balances the heat flux into the ledge surface from the metal. Presumably Sulmont and Hudault adjusted the metal–ledge heat transfer coefficient in some ad hoc fashion in the computations leading to these figures, since the model does not compute the melt flows.

Arita and coworkers[18] have examined heat transfer from bath or metal to the ledge. They used measurements of ledge profile, metal temperature, and

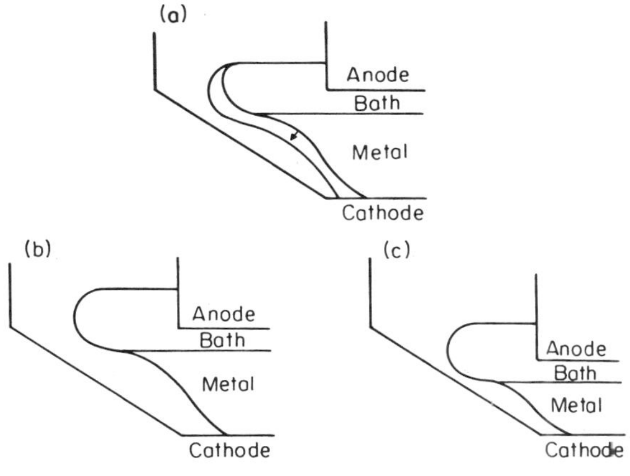

FIG. 8.7. (*a*) Melting of the ledge computed by a thermal model for a 2% increase in amperage. (*b*), (*c*) Effect of metal height (30 and 20 cm) on shape of the ledge. Reprinted with permission from *Light Metals 1978*, John J. Miller, ed., The Metallurgical Society of AIME, Warrendale, PA, 1978.

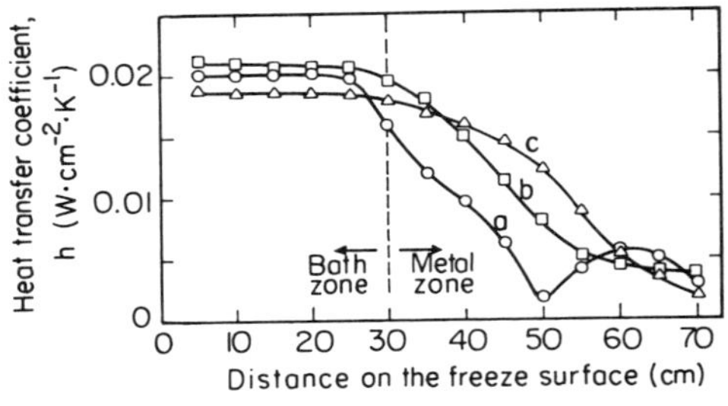

FIG. 8.8. Heat transfer coefficients from melt to freeze along the ledge. Reprinted with permission from *Light Metals 1978*, John J. Miller, ed., The Metallurgical Society of AIME, Warrendale, PA, 1978.

temperature in the cell wall (behind the ledge) carried out in an actual cell. These measurements were combined with a finite element computation of heat conduction in the ledge assuming the ledge surface exposed to the melt was at the bath liquidus temperature. From the computed temperature field in the ledge the heat flux at the melt–ledge interface could be computed. Dividing this flux by the temperature difference between melt and ledge surface yielded a heat transfer coefficient.

Figure 8.8 shows the heat transfer coefficient obtained in this way for a cell fed alumina from the side. The heat transfer coefficient between metal and ledge is lower than that between bath and ledge. This is anomalous, since the velocities in the metal are typically higher than those in the bath; furthermore, the thermal conductivity of bath is lower than that of metal. The explanation provided by Arita and coworkers is that side-feed cells typically have a high alumina level in the ledge (or in a film of molten electrolyte between ledge and metal), which results in a high liquidus temperature at this point. Consequently division of the metal–ledge heat flux by a temperature difference based on the bulk electrolyte liquidus yields a heat transfer coefficient that is spuriously low. An alternative explanation is that undissolved alumina at the metal–ledge interface interferes with heat transfer. Whatever the explanation for the low heat transfer coefficient, Arita and coworkers were essentially determining melt–ledge heat fluxes. Their heat transfer coefficients are therefore correct in the sense that they should be

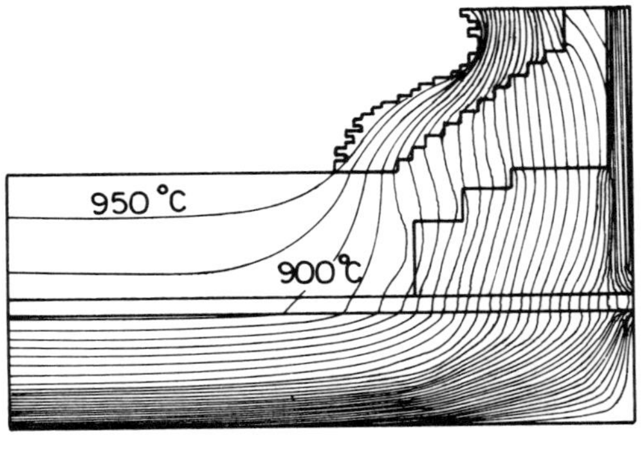

FIG. 8.9. Computed isotherms in a half section of the cell. Reprinted with permission from *Light Metals 1978*, John J. Miller, ed., The Metallurgical Society of AIME, Warrendale, PA, 1978.

used if the temperature difference between melt and ledge is that appropriate for the bulk electrolyte liquidus and melt temperature. Figure 8.9 shows the computed isotherms in a half section of the cell (the center plane of the cell is to the left in the figure). These isotherms were computed by Arita and coworkers using the heat transfer coefficient from Figure 8.8 by means of a thermal model similar to that of Sulmont and Hudault.

8.4. MODELS FOR FLUID FLOW AND INTERFACE SHAPE IN HALL–HÉROULT CELLS

Both the molten salt electrolyte (bath) and the metal are in motion in Hall–Héroult cells. In the case of the former melt, motion is due to a combination of electromagnetic forces and a gas lift effect brought about by the gas evolved at the anodes. The electromagnetic forces arise from the interaction of the magnetic field and the electric current within the bath; since the latter is predominantly in a downward direction and the electromagnetic force acts perpendicularly to the current vector, the electromagnetic force acts predominantly in a horizontal direction in the bath. In contrast, the forces due to the gas lift effect are vertical. They can be regarded as being due to a reduced effective density in those regions of the bath that contain bubbles. The bath in these regions therefore rises, while, to maintain continuity, bath in regions free of bubbles sinks with horizontal flows connecting the two regions.

Within the metal, the flow is driven by electromagnetic forces. Flows due to natural convection (arising from thermal gradients, or concentration gradients in the case of bath) are regarded as negligible in both melts, as are flows due to surface tension gradients.

As we have seen, electromagnetic forces act within the aluminum contained in a Hall–Héroult cell. These forces result in a motion of the aluminum that is predominantly in the horizontal direction and, at steady state, are balanced by (1) the inertia of the aluminum; (2) the drag force exerted by the sides of the cell, that is, the ledge, on the aluminum (horizontal transport of momentum); (3) the drag forces exerted by the cell bottom and bath on the aluminum (vertical momentum transport); and (4) forces arising from pressure gradients. Until recently most mathematical models (e.g., those of Lympany and coworkers[13, 14]) have assumed that item 3 could be treated as negligible. The fluid flow problem then becomes two-dimensional and is solved by standard computational techniques similar to those described in Chapter 6 of this book, with allowance made for turbulence, by means of the k–ε model, for example.

Within the molten salt electrolyte, motion is due to a combination of electromagnetic forces and the effect of gas generated at the anode. Again, at steady state, these forces are balanced by inertia, horizontal momentum transport, vertical momentum transport, and pressure forces. The models just referred to neglected vertical momentum transport as the price of obtaining a two-dimensional solution to the bath flow problem. Recently Moreau and

Evans[16] have shown that vertical transport of momentum cannot be neglected (i.e., that item 3 is significant) but that horizontal transport of momentum and inertia are negligible (i.e., that items 1 and 2 are insignificant). Furthermore, these investigators also showed that, in a cell with prebaked anodes, flow of bath in the channels between anodes was important and had a great impact on the circulation pattern. This flow in the channels between anodes had been neglected in previous models. The model of Moreau and Evans is summarized here.

First a shallow water approximation is made for each of the two melts; it is assumed that flow in each liquid can be described by its horizontal velocity averaged over the thickness of the liquid at the horizontal position in question.

$$\langle \mathbf{u}_{\mathrm{H}} \rangle = 1/H \int_{-\mathrm{H}}^{0} \mathbf{u}_{\mathrm{H}} dz \tag{7}$$

where $\langle \mathbf{u}_{\mathrm{H}} \rangle$ is the average horizontal velocity,
z is the vertical coordinate, and
H is the depth of the liquid.

The continuity and Navier–Stokes equations for such two-dimensional flow becomes

$$\nabla_{\mathrm{H}} \cdot \langle \mathbf{u}_{\mathrm{H}} \rangle = 0 \tag{8}$$

and

$$\rho \langle (\mathbf{u}_{\mathrm{H}} \cdot \nabla_{\mathrm{H}}) \mathbf{u}_{\mathrm{H}} \rangle = -\nabla_{\mathrm{H}} \langle p \rangle + \langle (\mathbf{j} \times \mathbf{B})_{\mathrm{H}} \rangle + \frac{\partial}{\partial x_{\mathrm{i}}} \left(\eta_{\mathrm{e}} \frac{\partial}{\partial x_{\mathrm{i}}} \langle \mathbf{u}_{\mathrm{H}} \rangle \right) - \frac{\boldsymbol{\tau}}{H} \tag{9}$$

In these equations boldface indicates vectors; the angle brackets indicate vertical averaging; x and y are the coordinates in the long and short horizontal directions, respectively; p is pressure; η_e is the effective viscosity (allowing for turbulence), and $\boldsymbol{\tau}$ is the shear stress (drag) exerted by the bounding horizontal surfaces on the liquid:

$$\nabla_{\mathrm{H}} \equiv \frac{\partial}{\partial x}, \quad \frac{\partial}{\partial y}, \quad 0 \tag{10}$$

In Eq. (9) the term on the left represents inertia. The first term on the right represents pressure forces, the second the electromagnetic force, the third and fourth the horizontal transport of momentum, and the last the drag averaged over the depth of the layer. Consider a cell of characteristic horizontal dimension (e.g., width) L. To appreciate the consequences of the small value of H/L, let us rewrite Eq. (9) in dimensionless form by introducing the following dimensionless variables:

$$P = \frac{p}{\mu j_0^2 L^2}, \qquad V = \left[\frac{\mu j_0^2 LH}{\rho}\right]^{1/2}$$

$$X = \frac{x}{L}, \qquad Y = \frac{y}{L}$$

$$\mathbf{U} = \frac{\langle \mathbf{u} \rangle}{V}, \qquad \mathbf{T} = \frac{\tau}{\rho V^2},$$

and

$$\mathbf{F} = \frac{\langle (\mathbf{j} \times \mathbf{B}) \rangle}{\mu j_0^2 L}$$

where μ is the magnetic permeability of the liquid,
j_0 is a representative current density (0.6 A/cm^2), and
ρ is the density of the liquid.

In the terms of these dimensionless variables, Eq. (9) becomes

$$\frac{H}{L}(\mathbf{U} \cdot \nabla_{\mathrm{H}})\mathbf{U} = -\nabla_{\mathrm{H}} P + \mathbf{F} + \frac{H}{L}\frac{\eta}{\rho V L}\frac{\partial}{\partial X}\left(\frac{\eta_{\mathrm{e}}}{\eta}\frac{\partial \mathbf{U}}{\partial X}\right) + \frac{H}{L}\frac{\eta}{\rho V L}\frac{\partial}{\partial Y}\left(\frac{\eta_{\mathrm{e}}}{\eta}\frac{\partial \mathbf{U}}{\partial Y}\right) - \mathbf{T} \tag{11}$$

For typical cells H/L is of the order of 10^{-2}, V is of the order 0 1 m/s, and $\eta/\rho VL$ is of the order 10^{-5}. We see, then, that the coefficients of the term on the left (inertial term) and the third and fourth term on the right (horizontal momentum transport) make these terms negligible. The fluid flow equations then take on the simple form

$$\nabla_{\mathrm{H}} \cdot \mathbf{U} = 0 \tag{12}$$

$$\mathbf{T} = \mathbf{F} - \nabla_{\mathrm{H}} P \tag{13}$$

If we further assume (for reasons discussed by Moreau and Evans[16]) that the drag on each liquid is proportional to the velocity of the liquid, we obtain

$$\kappa \mathbf{U} = \mathbf{F} - \nabla_{\mathrm{H}} P \tag{14}$$

where κ is a drag coefficient.

It should be noted that Eqs. (12) and (14) are a set of *linear* equations for the two unknowns **U** and P, and we can therefore anticipate a straightforward solution. Together with the stipulation of a boundary condition (that the normal component of **U** is zero at the sides and end of the cell) and the specification of the electromagnetic driving force **F**, this completes the mathematical formulation of equations describing flow in the aluminum.

The flow in the bath, however, is complicated by two effects: (1) the gas bubbles generated at the anode cause some flow of bath, and (2) the channels between anodes and around the periphery of the cell will affect the flow. Moreau and Evans neglected effect 1 but made allowance for effect 2. The importance of the channels arises from the fact that there are negligible current densities in these channels and therefore negligible electromagnetic forces. This contrasts with the regions of the bath underneath the anodes where there are electromagnetic forces acting inward toward the cell center. The channels therefore serve as return paths for the bath that is driven inward underneath the anodes.

Moreau and Evans divided the channels into two groups. The narrow channels between anodes on one (or the other) side of the cell were considered so narrow that negligible bath flowed along them; in other words, from the viewpoint of bath flow, these channels were ignored. The wide channels between the two rows of anodes (in a modern cell with center feeding) and between the anodes and the side of the cell were examined, and it was shown that these channels could sustain only negligible pressure gradients. For the bath, then, Eqs. (12) and (14) are solved with the boundary condition that P is constant along the wide channels.

It is worthwhile contrasting the flows in the two melts, since they are both obtained from the solution of the same pair of equations. For the aluminum we have the boundary condition being the specification of the normal velocity at the cell walls, but the pressure is not specified at the walls. For the bath, the boundary condition on the flow underneath the anodes is the specification of pressure (a constant) at the edge of anodes adjacent to wide channels, but no velocity is specified at these edges, permitting an interchange of bath between the region underneath the anodes and the channels.

As presented so far, the model of Moreau and Evans is very general and could employ values of the electromagnetic driving force **F** obtained by numerical computation (as described in Section 8.2), by analytical approximation, or even from experimental measurements. Moreau and Evans chose to illustrate their approach by considering an idealized cell. In this idealized cell the magnetic fields were considered to arise only from current within the bath and metal. Furthermore, it was assumed that the current in the bath was purely vertical. For the aluminum it was assumed that there was negligible current flow in the long horizontal direction and that the current density in the short horizontal direction and the vertical current density were both linear functions of position. With these simplifying assumptions the electromagnetic driving force can be obtained analytically. It is expressed in terms of a function G:

$$\mathbf{F} = \mathbf{j} \times \mathbf{B} = -\mu j_0^2 C \nabla_H G \tag{15}$$

where C is a dimensionless parameter representing the influence of horizontal currents (1 for no horizontal currents and declining for increased horizontal current). G is given by classical analytical solutions to the equation

$$\nabla_H^2 G = \begin{Bmatrix} 1 \text{ inside} \\ 0 \text{ outside} \end{Bmatrix} \begin{matrix} \text{the aluminum pool or in} \\ \text{the bath under the anodes} \end{matrix} \tag{16}$$

Substituting Eqs. (15) and (16) in Eq. (14) yields

$$\kappa \mathbf{U} = -\nabla_H(P + G) \tag{17}$$

Taking the divergence of Eq. (17) and using Eq. (16) yields an equation for pressure distribution in the bath ($C = 1$):

$$\nabla_H^2 P = -1 \tag{18}$$

for which we use the boundary condition $P = 0$ (an arbitrary constant) along the edge of the rectangle representing the wide channels. With this boundary condition, Eq. (16) also yields an analytical solution. Knowing the distribution of P and G in the bath then gives the velocity by application of Eq. (17).

With the driving force of Eq. (15) and the boundary condition that the normal velocity be zero at the sides of the cell, the analytical solution of these equations for the aluminum yields $\mathbf{U} = 0$ everywhere throughout the metal! This is a consequence of the idealizations; in real cells the metal moves with velocities of the order of 10 cm/s, and we deduce that this is a consequence of the magnetic field due to currents external to the melts, to nonlinear variation of current density in the metal, or (most likely) to both.

Equation (17) now becomes

$$P + G = \text{constant} \tag{19}$$

for the aluminum, permitting calculation of the pressure distribution in the metal pool. Finally, knowledge of the pressure distribution in the aluminum and bath yields the shape of the interface between the two melts from the continuity of pressure across the interface (see Moreau and Evans[16] for details).

Figure 8.10 shows some results of this model. Figure 8.10*a* shows the calculated bath circulation pattern (streamlines) for a cell without a wide central channel. Such a cell would be either one with a self-baking anode or an older cell where there is only a small gap between the two rows of anodes. The cell is seen from above, and only the northeast quadrant of the cell is depicted, the other

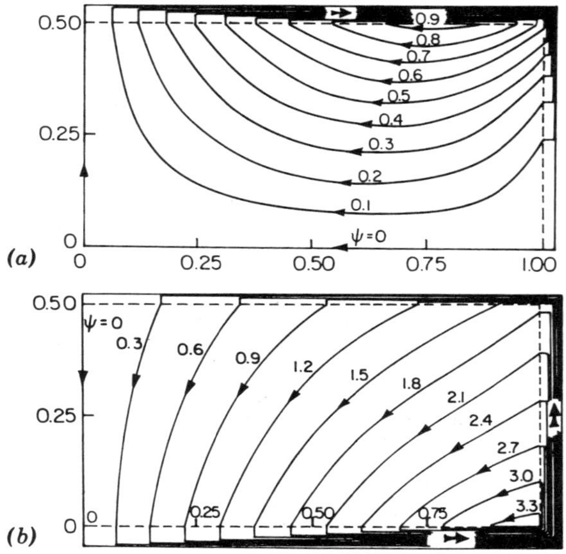

FIG. 8.10. Streamlines for electrolyte flow under the anodes, computed by the model of Moreau and Evans (*a*) for a cell without a central channel and (*b*) for a cell with a central channel.[16] From *J. Electrochem. Soc.*, **131**, 2255 (1984), Reprinted by permission of the publisher, The Electrochemical Society, Inc.

quadrants being symmetrical. The stream function is dimensionless (which makes it unnecessary to specify parameters such as j_0 or κ), and it will be recalled that the slope of the stream function along a line perpendicular to a streamline is proportional to the velocity along the streamline. We see that the flow is directed under the anodes from the short sides of the cell and that (except in the channel around the cell periphery) velocity is highest at a point near the upper right.

Figure 8.10*b* is for a cell with a wide central channel, and we see that the circulation pattern is quite different. The bath is flowing from both the long and the short sides of the cell in a direction roughly toward the center. The flow then returns outward along the outer channel. Note that the flow of bath along the long side of the cell is opposite to that in Figure 8.10*a*. Furthermore, the stream function gradients are much greater in Figure 8.10*b*, indicating much higher velocities for the cell with a center channel.

Figure 8.11a shows the bath–metal interface shape for a cell without a central channel. The whole interface is depicted, rather than one quadrant. The numbers give the interface position at the point with respect to the mean interface position (that which would exist if the interface were flat). The shape is calculated for $C = 0.5$ (significant horizontal currents in the aluminum). The interface is seen to have a saddle shape with the center being the lowest point. The shape may be explained as follows. Starting at one corner of the cell and moving along the long side, pressure in the bath remains constant (since this is a

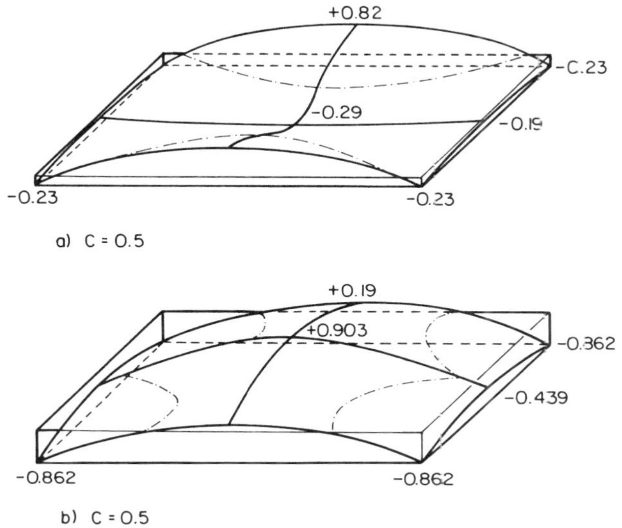

FIG. 8.11. Computed bath–metal interface shape (*a*) for a cell without a central channel and (*b*) for a cell with a central channel.

wide channel), but components of the electromagnetic force acting in the long direction cause the pressure in the metal to increase, causing the metal to bulge upward into the bath. Reaching the middle of the long side, our observer turns inward toward the cell center. Now the pressure increases in both the bath and the metal. The former increase is greater than the latter because (1) while the aluminum is stationary the bath is traveling in the opposite direction to our observer, causing an extra pressure increase in the bath, and (2) in the metal the inward electromagnetic forces due to vertical current and horizontal magnetic field are diminished as the horizontal current drains current from the vertical direction. The observer therefore experiences a greater pressure increase in the bath than in the metal, and the bath eventually reaches the higher pressure, bulging down into the metal.

The overpressure in the bath at the center of the cell would be relieved by a center channel, as depicted in Figure 8.11*b*. Now, because of this relief of the overpressure in the bath, the interface between metal and bath bulges upward, even at the center of the cell. Note that the total deformation of the interface (valley to peak distance) is much greater in the presence of the center channel.

The approach of Moreau and Evans to fluid flow in the cell is significant in that it obviates the need for consideration of turbulence except inasmuch as turbulence affects the drag coefficient κ. The approach may be further advanced by allowing for exchange of momentum between the two liquid layers. **T** in Eq. (13) would then depend on velocity in both liquids, rather than just one. Another improvement might be the incorporation of the effect of gas bubbles into the bath flow. These authors illustrated the application of their approach by considering an idealized cell where simplifying assumptions for the magnetic field and current distribution yielded analytical expression for the electromagnetic driving forces. An obvious next step would be the incorporation of more realistic electromagnetic forces (perhaps obtained by numerical calculation); indeed, these authors presented a second-order improvement in the form of a perturbation on the first model to treat the case of a nonlinear current density in the metal. In this second-order improvement, the metal was no longer found to be stationary.

8.5. PHYSICAL MODELING OF HALL–HÉROULT CELLS

By now the reader will be aware that the mathematical modeling of Hall–Héroult cells is a difficult task. For example, the flow of the molten salt electrolyte is three-dimensional, turbulent, two-phase, and driven in part by electromagnetic forces. Consequently, there have been many attempts to use physical models to gain insight into cell behavior. Some examples are treated here.

Dernedde and Cambridge[19] constructed a physical model (Figure 8.12) aimed at studying gas-induced circulation of the electrolyte. The electrolyte was simulated by water and the metal by a mixture of silicone oil with perchloroethylene. The model was a vertical slice with electrolytic generation of gas at the Soderberg anode simulated by forcing air through porous alundum plates representing the bottom of the anode. Movable curved metal plates represented the ledge at the side of the cell. The choice of water and organic phase resulted in a small density difference between the two liquids (as in the real cell) and kinematic viscosities that were comparable to those in the real cell, although the interfacial tension between the bath and the metal was different in the model than in the real case.

Dernedde and Cambridge observed waves at the interface between the two

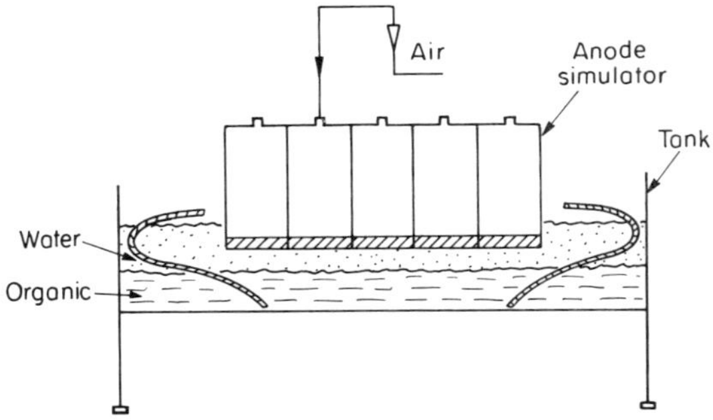

FIG. 8.12. Physical model for studying flow driven by bubbles and natural convection in a Hall–Héroult cell. Reprinted with permission from *Light Metals* 1975, R. Rentsch, ed., The Metallurgical Society of AIME, New York, 1975.

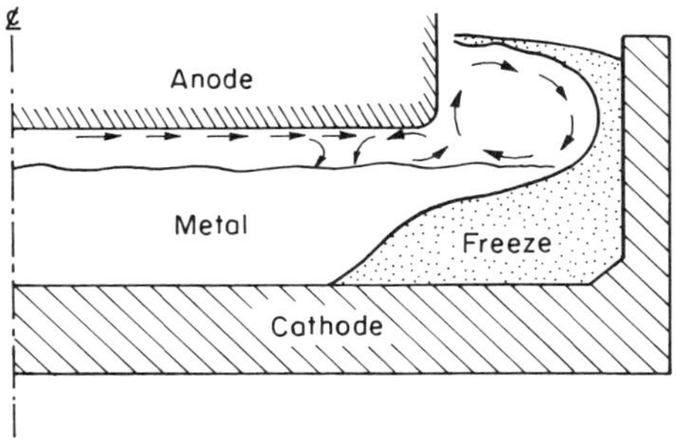

FIG. 8.13. Flow patterns driven by bubbles in a Hall–Héroult cell suggested by physical modeling. Reprinted with permission from *Light Metals 1975*, R. Rentsch, ed., The Metallurgical Society of AIME, New York, 1975.

liquids similar to those observed in real cells. These waves became smaller in magnitude as the metal strips representing the ledge (or freeze) were moved inward. With the interfacial area restricted to an area less than the area beneath the anode, the interfacial waves were almost eliminated. Figure 8.13 shows the circulation pattern in a cell predicted by these authors on the basis of their modeling work.

In a separate series of experiments Dernedde and Cambridge measured velocities in the region between the anode and the side of the cell. This was done by taking movie films of the circulation of polystyrene spheres of neutral density. The geometry (anode to side wall distance and anode immersion), gas flow rate, and physical properties were varied in these experiments, and an empirical correlation was obtained for the speed of the "electrolyte" as a function of these parameters. These investigators coupled this correlation to a correlation for heat transfer to a flat plate immersed in a moving liquid to obtain predictions of freeze thickness in a real cell. These predictions are shown in Figure 8.14, where the horizontal axis is the anode to side wall distance.

The predictions are for both a Soderberg anode (where all the anode gas passes through the region between the side of the anode and the side wall) and prebaked anodes (where much of the gas passes through gaps between anodes, reducing the gas flow between anode and side wall). The predictions are consistent with expectations; the circulation at the side wall is more vigorous in the Soderberg case, which increases the heat transfer to the freeze and thereby reduces its thickness. Furthermore, the circulation is enhanced by a deeper immersion of the anode (which increases the gas lift effect), and therefore a deeper immersion of the anode results in a thinner freeze.

Maintaining a sufficiently thick freeze is the procedure used to prevent attack of the cell side wall by the bath and is therefore an important aspect of cell behavior. Solheim and Thonstad[20] have followed the lead of Dernedde and Cambridge in addressing this issue by means of physical models. Solheim and Thonsted used a slice model that was constructed as a full-scale transverse section of an aluminum cell from the center line to the side wall (see Figure 8.15). The width of the section was 0.7 m, and one liquid (molten biphenyl at 75°C) was used. The kinematic viscosity and the Prandtl and Grashof numbers for molten

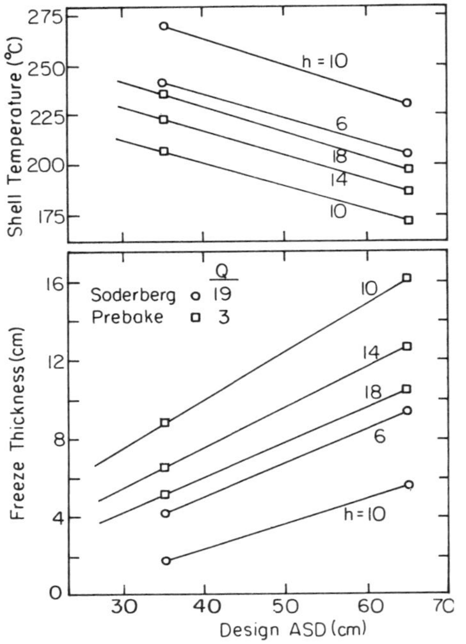

FIG. 8.14. Heat transfer coefficients between freeze (ledge) and bath predicted from velocity measurements on physical models. Reprinted with permission from *Light Metals 1975*, R. Rentsch, ed., The Metallurgical Society of AIME, New York, 1975.

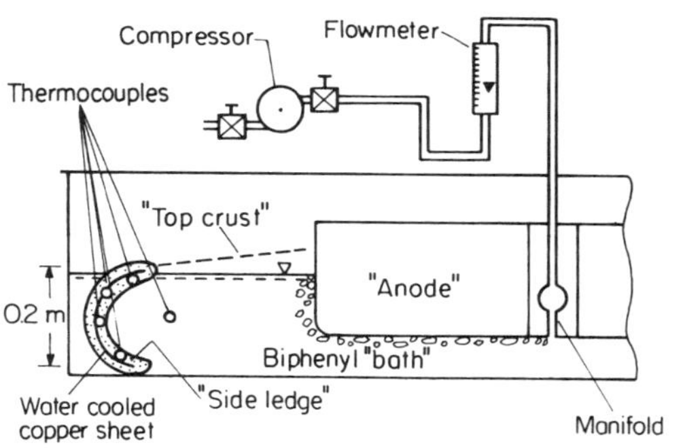

FIG. 8.15. Physical model constructed by Solheim and Thonstad.[20] Reprinted with permission from *Light Metals 1983*, E. M. Adkins, ed., The Metallurgical Society of AIME, Warrendale, PA, 1983.

biphenyl are similar to those of the bath in a real cell. A water-cooled copper sheet, onto which biphenyl was frozen, was used to simulate the frozen ledge (freeze) at the side of the cell. Thermocouples within the frozen biphenyl and copper enabled the determination of the local heat flux from a knowledge of the thermal conductivity of biphenyl. The local heat transfer coefficient could then be determined by dividing the heat flux by the temperature difference between the molten biphenyl and the freezing point of biphenyl.

These investigators found that their results could be correlated by an equation of the form

$$h = (\text{constant})\,(\text{GF} \cdot \text{AI}/\text{ALD})^k \tag{20}$$

where h = local heat transfer coefficient,
GF = anode gas flow rate,
AI = anode immersion, and
ALD = anode to ledge distance.

Results appear for three vertical positions in Figure 8.16. These heat transfer coefficients are in the absence of anything simulating a frozen crust on the top of the cell or the metal phase. Decreases in heat transfer coefficient were observed when these were simulated. With correction for these effects and for the different physical properties of the bath in the industrial cell, these investigators arrived at the correlation for the Hall–Héroult cell:

$$h_b = 236\,(\text{GF} \cdot 10^4 \cdot \text{AI}/\text{ALD})^{0.42} \tag{21}$$

This correlation (in SI units) is for the average heat transfer coefficient over the vertical extent of the freeze and is thought to be precise to +/− 20%. The model of Solheim and Thonstad (and also that of Dernedde and Cambridge) does not

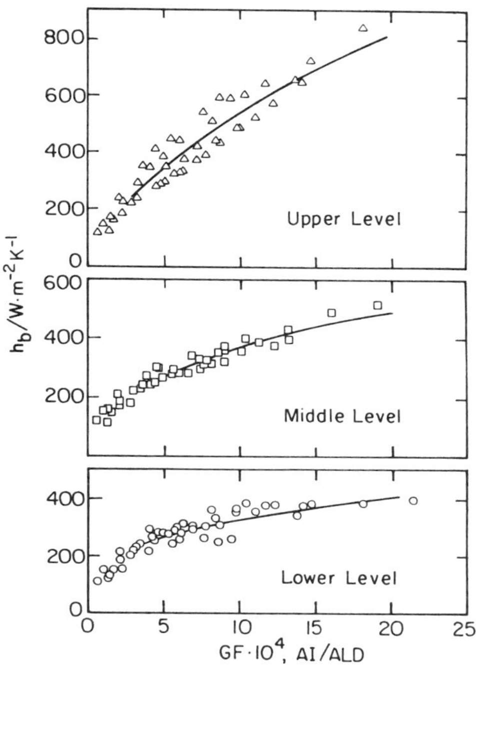

FIG. 8.16. Heat transfer coefficients measured by Solheim and Thonstad as a function of their parameter GF·AI/ALD
where GF = gas flow rate per unit length of anode periphery (m^2s^{-1}),
AI = anode immersion (m),
ALD = anode ledge distance (m).[20]
Reprinted with permission from *Light Metals 1983*, E. M. Adkins, Warrendale, PA, 1983.

include the horizontal flow driven by electromagnetic forces that is experienced by the bath in a real cell. Solheim and Thonstad presented an approach that enables the correction of the heat transfer coefficient for this flow if the horizontal velocity is known; their correction is of the order of a 20% increase.

Physical modeling of the gas-driven circulation in Hall–Héroult cells has also been carried out by Fortin and coworkers.[21] Figure 8.17 shows the model used by these investigators. It is seen that only one liquid (water) is used (representing the bath), and provision is made for pumping this water through the model independently of the circulation due to gas bubbles. The emphasis of this work was on the shape and size of the anode gas bubbles, and video cameras recorded these bubbles by means of a mirror mounted underneath the model. Provision was also made for tilting the anode from the horizontal. The effects of current density (gas flow rate), anode–cathode distance, anode tilt, and electrolyte velocity on gas layer geometry, anode coverage by gas, bubble velocity, and gas release frequency were investigated. Under some conditions, bubbles up to 2 cm thick were produced.

The physical modeling of electromagnetically driven flows has been carried out by Lee and Evans.[22] One objective of their investigation was to obtain measurements for comparison with mathematical models for electromagnetically driven flow in Hall–Héroult cells. A second objective was to physically model cell behavior under circumstances in which mathematical modeling would be difficult. Figure 8.18 is a half section of the model used by Lee and Evans; the model could be shortened by removal of end "anodes" and insertion of plastic walls. The model was imperfect in that it contained Wood's metal (melting point, 70°C), which moves only under the influence of electromagnetic forces, with no

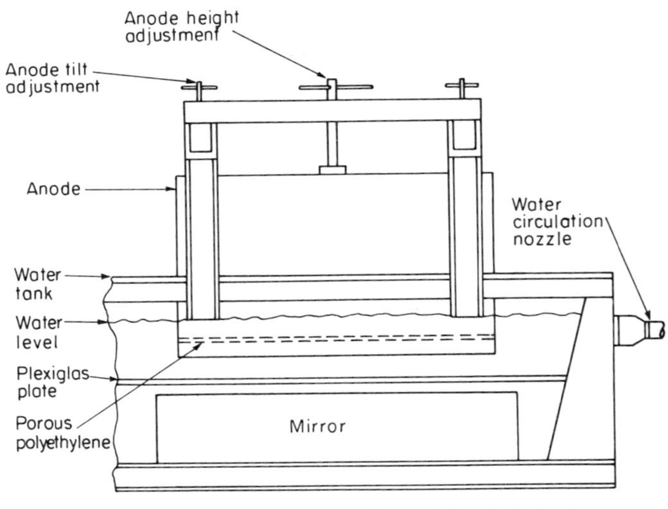

FIG. 8.17. Schematic diagram of water model used for studying bubble phenomena in Hall cells. Reprinted with permission from *Light Metals 1984*, J. P. McGeer, ed., The Metallurgical Society of AIME, Warrendale, PA, 1984.

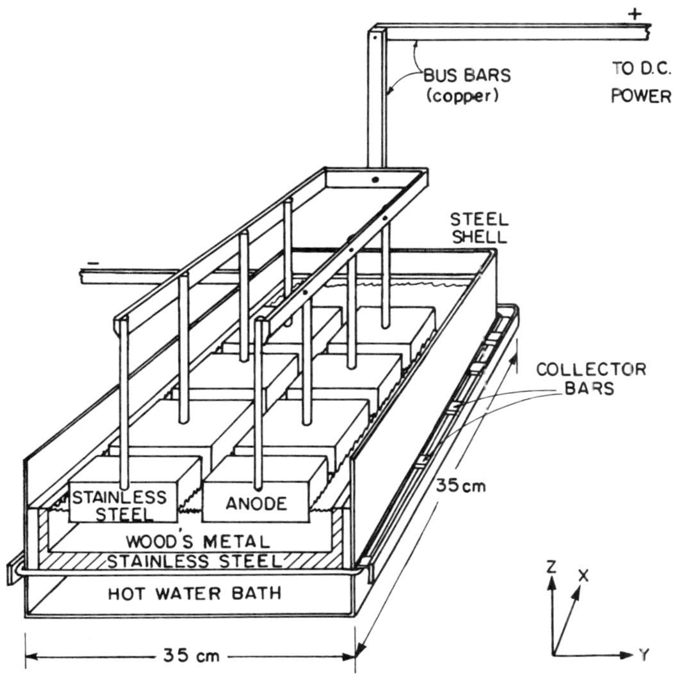

FIG. 8.18. Sectional drawing of physical model used for studying electromagnetically driven flow in Hall–Héroult cells. Reprinted with permission from *Light Metals 1985*, H. O. Bohner, ed., The Metallurgical Society of AIME, Warrendale, PA, 1985.

evolution of gas at the "anodes." Wood's metal was selected as the model fluid because of its high electrical conductivity (compared, say, to aqueous solutions) and low melting point; the latter makes velocity measurements feasible by the electromagnetic technique of Vives (described in Chapter 4), where motion of a melt in the field of a small permanent magnet generated a voltage difference between two electrical leads inserted alongside the magnet. The model was driven by a dc power supply at current densities comparable with those of a real cell. The Wood's metal was kept molten by a hot water jacket and plastic foam insulation.

In collaboration with Vives, the authors[23] carried out calibration experiments for the Vives electromagnetic velocity probe in Wood's metal. The calibration curves appear in Figure 8.19. A difficulty with the use of the probe in the Hall–Héroult cell model was that the probe was responsive to two voltage gradients. The first of these was the voltage gradient produced by metal motion in the probe's magnetic field, and the second was the voltage gradient resulting from passage of current through the model. The former was only a few percent of the latter. This difficulty was overcome by a current interruption technique. On switching off the dc power supply, the latter voltage gradient disappeared immediately, while the former diminished over a period of several tens of

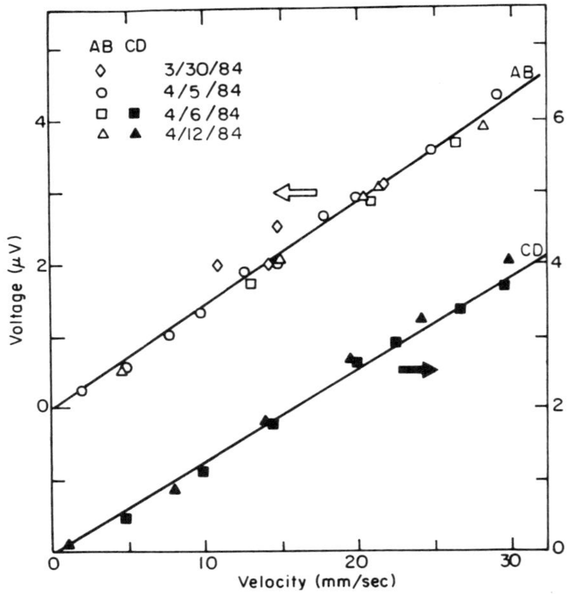

FIG. 8.19. Calibration curves for the magnetohydrodynamic velocity probe of Vives in Wood's metal. Reprinted with permission from *Light Metals 1985*, H. O. Bohner, ed., The Metallurgical Society of AIME, Warrendale, PA, 1985.

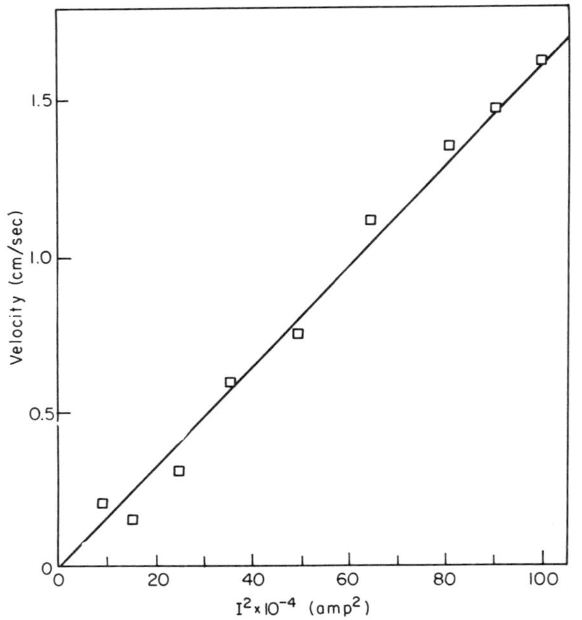

FIG. 8.20. Velocity of Wood's metal at a point in the physical model of the Hall–Héroult cell versus the square of the current. Reprinted with permission from *Light Metals 1985*, H. O. Bohner, ed., The Metallurgical Society of AIME, Warrendale, PA, 1985.

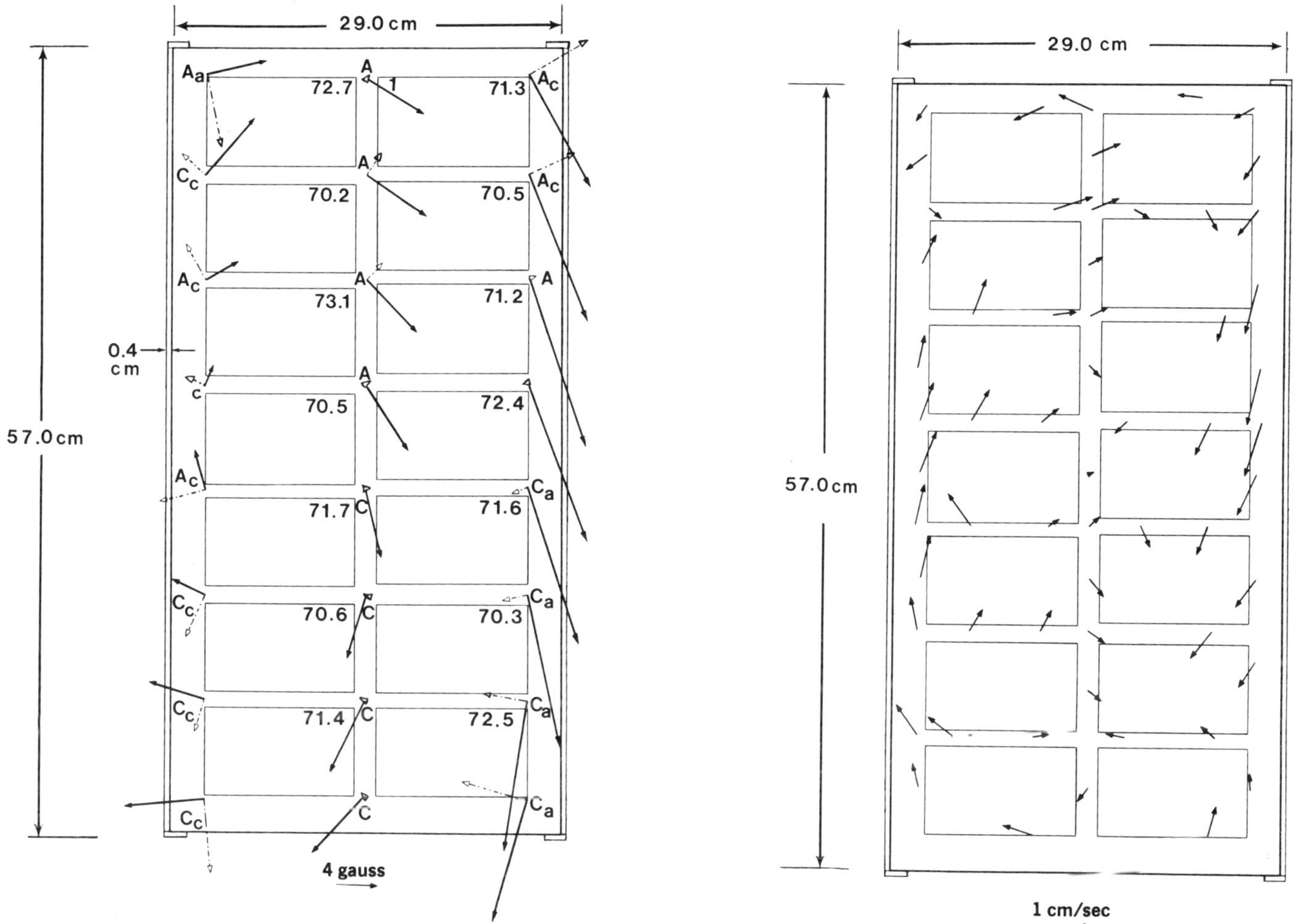

FIG. 8.21. Velocity vectors and magnetic fields as measured in the physical model for electromagnetically driven flow in Hall–Héroult cells.[24] Reprinted with permission from *Light Metals 1987*, edited by R. D. Zabreznik, ed., The Metallurgical Society of AIME, Warrendale, PA, 1987.

seconds as the Wood's metal (weighing 55 kg) was gradually brought to rest by drag forces. A plot of voltage versus time could therefore be extrapolated backwards to zero time to give the velocity at the instant the power supply is turned off.

Figure 8.20 shows the velocity measured at a point in the Wood's metal versus the square of the current through the model. The observed proportionality is consistent with the work of Moreau and Evans,[16] discussed is Section 8.4. Those investigators had concluded that flow in a Hall–Héroult cell is largely determined by a balance between drag (proportional to the velocity) and the driving force (proportional to current squared), with differences between the two resulting in pressure gradients. A proportionality between velocity and the square of the current was therefore to be expected.

Figure 8.21 shows the horizontal magnetic field vectors and the velocity vectors recently measured by Banerjee and Evans in the same apparatus for a case in which 14 anodes are present. The model is seen from above, and the arrow lengths are proportional to velocity and magnetic field strength. The magnitudes of the velocity vectors were approximately one order of magnitude lower than in a real cell (as expected for the lower magnetic field). The distribution of current among the anodes is given by the numbers within each anode. Broken arrows indicate the vertical component of the magnetic field and the convention used in Figure 8.21 is that these arrows are drawn clockwise from the solid arrows (representing the horizontal field) if the vertical field is downward. The letters appearing on the figure showing the magnetic field results (A, C, a, and c) indicate whether the torque resulting from the interaction of the horizontal magnetic field and vertical current is anticlockwise or clockwise (capital letters) and the direction of the torque from the vertical field and horizontal current (lowercase letters). The dominance of clockwise torques should be noted and results in the predominantly clockwise swirl revealed in the velocity measurements.

8.6. CONCLUDING REMARKS

The role played by the Hall–Héroult cell in the aluminum industry is important, and cell performance, particularly specific energy consumption, is of great significance to the companies in that industry. A brief description of these cells and of the physicochemical phenomena occurring within them was provided here.

Mathematical models have been developed to describe electrical conduction, heat generation and transport, magnetic fields and electromagnetic forces, melt flow, mass transport, and bath–metal interface deformation. Because of the complexity of the phenomena involved, the models described in the literature all involve approximations to various extents; furthermore, none of these models is complete in the sense of incorporating all the important phenomena.

Physical models have been oriented toward studying melt flow driven by anode gas bubbles or by electromagnetic forces. These models are also piecemeal in that a complete physical model would incorporate both driving forces (and perhaps also heat transport).

The field of Hall–Héroult cell modeling is an active one. From papers presented at meetings and from personal contacts, the authors are aware of the fact that all the major aluminum companies in the United States and abroad have made substantial efforts in this direction. It is believed that these efforts are mandated by the increasing cost of electrical energy in many parts of the world and the enormous and readily calculable benefits resulting from improvements of as low as a few percent in cell performance. In recent years there has been an increasing interest in Hall–Héroult cell modeling in the academic community, notably at Trondheim, Grenoble, and Berkeley.

REFERENCES

1. N. Jarrett, in Electrochemistry Research Needs for Mineral and Primary Materials Processing, T. J. O'Keefe and J. W. Evans, eds., report of workshop at University of Missouri-Rolla, 1983.
2. K. Grjotheim, C. Krohn, M. Malinovsky, K. Matiasovsky, and J. Thonstad, *Aluminum Electrolysis*, Aluminium-Verlag, Düsseldorf, West Germany, 1977, chap. 9
3. B. Berge, K. Grjotheim, K. Krohn, R. Naeumann, and K. Torklep, *Light Metals 1975*, Metallurgical Society of AIME, New York, 1975, p. 37.
4. O. A. Asbjornsen, J. A. Andersen, and A. Larsen, *Light Metals 1979*, Metallurgical Society of AIME, Warrendale, PA, p. 517.
5. A. Furman, *Light Metals 1978*, Metallurgical Society of AIME, Warrendale, PA, 1978, p. 87.
6. T. Ohta and T. Matsushima, *Light Metals 1984*, Metallurgical Society of AIME, Warrendale, PA, 1984.
7. G. A. Solinas, *Light Metals 1978*, Metallurgical Society of AIME, Warrendale, PA, p. 15.
8. R. F. Robl, *Light Metals 1978*, Metallurgical Society of AIME, Warrendale, PA, p. 1.
9. E. D. Tarapore, *Light Metals 1979*, Metallurgical Society of AIME, Warrendale, PA, p. 541.
10. E. D. Tarapore, *Light Metals 1981*, Metallurgical Society of AIME, Warrendale, PA, p. 341.
11. J. W. Evans, Y. Zundelevich, and D. Sharma, *Met. Trans.*, **12B**, 343 (1981).
12. T. Arita and H. Ikeuchi, *Light Metals 1981*, Metallurgical Society of AIME, Warrendale, PA, p. 357.
13. S. D. Lympany and J. W. Evans, *Met. Trans.*, **14B**, 63 (1983).
14. S. D. Lympany, D. P. Ziegler, and J. W. Evans, *Light Metals 1983*, Metallurgical Society of AIME, Warrendale, PA, p. 507.
15. R. F. Robl, *Light Metals 1983*, Metallurgical Society of AIME, Warrendale, PA, p. 449.
16. R. Moreau and J. W. Evans, *J. Electrochem. Soc.*, **131**, 2251 (1984).
17. B. Sulmont and G. Hudault, *Light Metals 1978*, Metallurgical Society of AIME, Warrendale, PA, p. 73.
18. Y. Arita, N. Urata, and H. Ikeuchi, *Light Metals 1978*, Metallurgical Society of AIME, Warrendale, PA, p. 59.
19. E. Dernedde and E. L. Cambridge, *Light Metals 1975*, Metallurgical Society of AIME, New York, p. 111.
20. A. Solheim and J. Thonstad, *Light Metals 1983*, Metallurgical Society of AIME, Warrendale, PA, p. 425.

21. S. Fortin, M. Gerhardt, and A. J. Gesing, *Light Metals 1984*, Metallurgical Society of AIME, Warrendale, PA, p. 721.

22. H.-C. Lee and J. W. Evans, *Light Metals 1985*, Metallurgical Society of AIME, Warrendale, PA, p. 569.

23. H.-C. Lee, J. W. Evans, and C. Vives, *Met. Trans.*, **15B**, 734 (1984).

24. S. Banerjee and J. W. Evans, *Light Metals 1986*, Metallurgical Society, Inc., Warrendale, PA, p. 535.

INDEX